E.W. TAYLOR.

021-414-5472

SOCIETY FOR EXPERIMENTAL BIOLOGY
SEMINAR SERIES • 16

GILLS

GILLS

Edited by

D.F.HOULIHAN

Department of Zoology
University of Aberdeen

J.C.RANKIN

Department of Zoology
University College of North Wales, Bangor

and

T.J.SHUTTLEWORTH

Department of Biological Sciences
University of Exeter

CAMBRIDGE UNIVERSITY PRESS

Cambridge

London New York New Rochelle

Melbourne Sydney

Published by the Press Syndicate of the University of Cambridge
The Pitt Building, Trumpington Street, Cambridge CB2 1RP
32 East 57th Street, New York, NY 10022, USA
296 Beaconsfield Parade, Middle Park, Melbourne 3206, Australia

First published 1982

Printed in USA by
Vail Ballou Press, Inc.,
Binghamton, N.Y.

Library of Congress catalogue card number: 81-21778

British Library cataloguing in publication data

Gills. – (Society for Experimental Biology
seminar series; 16)

1. Gills – Congresses
I. Houlihan, D. F. II. Rankin, J. C.
III. Shuttleworth, T. J. IV. Series
591.1'2 QL846

ISBN 0 521 24083 2 hard covers

CONTENTS

CONTRIBUTORS

Bolis, L., Department of General Physiology, University of Messina, Sicily, Italy.
Bolt, S. R. L., Department of Oceanography, The University, Southampton SO9 5NH, UK.
Dawson, M. E., Department of Oceanography, The University, Southampton SO9 5NH, UK.
Evans, D. H., Department of Zoology, University of Florida, Gainsville, Florida 32611, USA.
Goldstein, L., Division of Biology and Medicine, Brown University, Providence, Rhode Island 02912, USA.
Hughes, G. M., Research Unit for Comparative Animal Respiration, The University, Woodland Road, Bristol BS8 1UG, UK.
Johansen, K., Department of Zoophysiology, University of Aarhus, DK-8000 Aarhus C., Denmark.
Kirschner, L. B., Department of Zoology, Washington State University, Pullman, Washington 99163, USA.
Laurent, P., CNRS Laboratoire de Physiologie comparée des Régulations, 23, Rue du Loess, 67037 Strasbourg-Cédex, France.
Lockwood, A. P. M., Department of Oceanography, The University, Southampton SO9 5NH, UK.
Mangum, C. P., Department of Biology, College of William and Mary, Williamsburg, Virginia 23185, USA.
Piiper, J., Max-Planck-Institut fur experimentelle Medizin, Abteilung Physiologie, Hermann-Rein-Strasse 3, D-3400 Gottingen, West Germany.
Randall, D., Department of Zoology, The University of British Columbia, 2075 Wesbrook Mall, Vancouver, B.C. V6T 1W5, Canada.
Rankin, J. C., Department of Animal Biology, University College of North Wales, Bangor, Gwynedd LL57 2UW Wales, UK.
Scheid, P., Max-Planck-Institut fur experimentelle Medizin, Abteilung Physiologie, Hermann-Rein-Strasse 3, D-3400 Gottingen, West Germany.
Stagg, R. M., Department of Biological Sciences, University of Exeter, Exeter EX4 4PS, UK.

PREFACE

This is probably the first volume which concentrates on the topic of gills. It consists of a series of reviews by people who have approached, from a number of different angles, the exchange processes that occur across gills. The chapters are arranged to cover structures of gills, the physical principles of gases, solute and water transfer across gills, and descriptions of these processes in invertebrates and vertebrates. There are also chapters on blood flow through gills, nitrogen excretion and the effects of pollutants.

The seminar series of the Society for Experimental Biology, of which this is the sixteenth volume, is intended primarily for advanced undergraduates and postgraduates. This volume is a record of the major papers given at a symposium of the Society in Durham University, held on 24–25 March 1981. We would like to thank the chairmen and other contributors to the Gills symposium who made it such an enjoyable and stimulating affair. We would also like to thank Dr J. Anstee who acted as local organiser of the conference of which this symposium formed a part.

June 1981 D.F.H., J.C.R., T.J.S.

SYMBOLS AND UNITS

Some different symbols and units have been used by the individual authors in this volume. The symbols used in respiratory physiology are mainly adapted from Dejours' *The Principles of Comparative Respiratory Physiology*. A glossary of terms on respiration and gas exchange is given by Bartels *et al*. (1973) *Journal of Applied Physiology,* **34,** 549–58. The following symbols are used in this volume:

a	arterial
v	venous
i	inspired
e	expired
K	Krogh permeation coefficient
P_{50}	partial pressure of oxygen at which 50% of the respiratory pigment is oxygenated
P	gas pressure (may be partial pressure)
$\dot{V}$	volume per unit time
$\dot{Q}$	blood flow per unit time

Other symbols are defined by individual contributors.

Equivalent units

1 mm Hg = 1 Torr = 0.133 kPa

G.M.HUGHES

An introduction to the study of gills*

As this volume is probably the first to concentrate on 'gills', perhaps it is appropriate to consider the origin of this word. The etymological dictionary suggests that it is of uncertain origin† but may be derived from an old Norse word (gjolnar) meaning 'whiskers of the mythical Fenris wolf', but at least six other nouns have the same spelling. One of these is used to describe clefts in hill formations in parts of northern England but has quite a separate origin; the spelling 'ghyll' was introduced by William Wordsworth. Thus it would appear that the view of a gill as an outpushing from the body wall of an animal is a very old one but originally had little to do with respiratory function. William Horman (1519) was perhaps the first to have this view when he wrote 'fysshes breth at theyr gylls'. 'That their Gills seem somewhat Analogous (as to their use) to Lungs' was suggested by Robert Boyle (1660).

During the more recent study of these organs the occurrence of 'non-whiskery' gills has become apparent. For example, blood and tracheal gills are well known among insects (Fig. 1) but so also is the concept of a 'physical gill' whereby a bubble of gas taken down by an aquatic insect serves not only as a store of oxygen but also as an organ into which oxygen diffuses when its partial pressure is greater in water than within the bubble (Ege, 1915; De Ruiter, Wolvekamp, van Tooren & Vlasblom, 1952; Rahn & Paganelli, 1968). Some insects do not need to visit the surface in order to replenish their bubbles of gas because the gas is trapped in a 'plastron' which is of constant volume and serves as a permanent physical gill (Thorpe, 1950; Crisp, 1964; Hinton, 1966). Thus gills have very often been viewed as respiratory structures but all extensions of the body wall are not necessarily concerned with gas exchange. Such a problem was clearly demonstrated with insect larvae by the use of protozoans as biological indicators since they aggregate at a particular oxygen tension (Fig. 2). Munro Fox (1921) demonstrated the respiratory function of the horns of *Simulium* pupae whereas

*Dedicated to the memory of Dr I. E. Gray of Duke University, Durham, North Carolina who helped my own introduction to gills in correspondence which began in 1958–9.

†The origin of the German word for a gill, Kieme, is also obscure, in fact in the sixteenth and seventeenth centuries gills were called Fischohren (fish ears!). Later Kiemen came perhaps from Kiefer (jaw) or Kimme meaning a notch or cleft.

the blood gills of *Chironomous* were not especially important. The same technique also showed that the gill-like extensions of some parasitic insect larvae may have a respiratory function (Fig. 2*b*) but others do not serve for the transfer of oxygen any more than other parts of the body wall (Thorpe, 1930). Thorpe (1932) also found a similar, but sometimes different, distribution for the regions of carbon dioxide release which were detected by dyes such as cresol red when added to the water containing the larvae.

Gills are known to have many functions. This is not surprising if one takes a comparative viewpoint as functions such as osmoregulation (Crustacea), loco-

Fig. 1. Diagrams to illustrate different types of insect gills and their function in gas exchange. (*a*) Tracheal gills of dragonfly larva. (*b*) Physical gill of water beetle, and diagram of a plastron, or permanent physical gill. (After Hughes & Kylstra, 1964.)

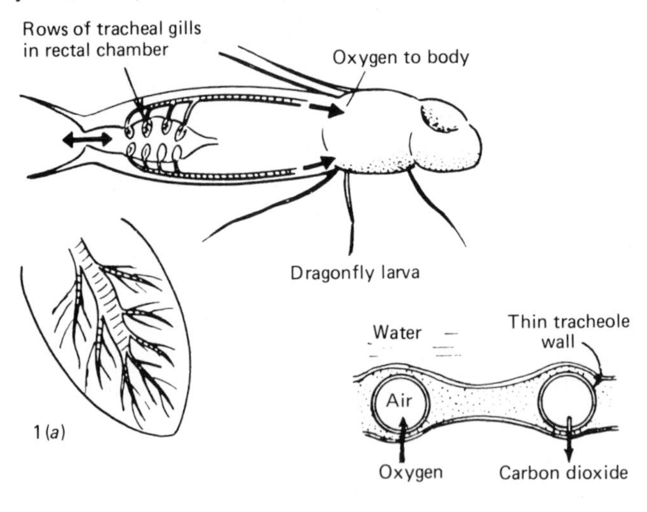

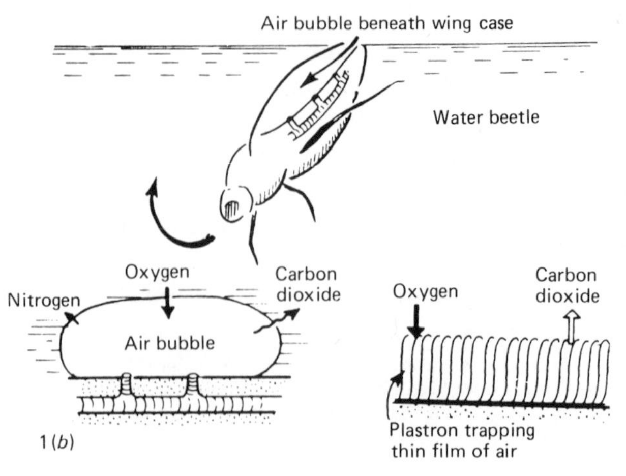

motion (Septibranchs) and feeding (tubicolous Polychaetes) occur in a number of animal groups. Among the Chordates, where gills are a diagnostic character, maintenance of a ciliary feeding current appears to have been their earliest function, persisting among present-day forms such as *Amphioxus* (Baskin & Detmers, 1976) and the ammocoetes larva of lampreys (Youson & Freeman, 1976). The presence of cilia (Fig. 4) on the gill filaments and secondary lamellae of the primitive actinopterygian fish, *Polypterus* (Hughes, 1980*a*) may be a legacy from this ancestral condition.

A characteristic feature of gills is the way that their structure and development varies according to the type of organism and its environment. An early example of this kind is illustrated in Fig. 3 where extensions of the body wall at the

Fig. 2. (*a*) Diagram to show patterns of aggregation of protozoan (*Bodo*) round pupa of *Simulium*. (After Fox, 1921.) (*b*) Parasitic larvae of (i) *Omorgus* (Ichneumonidae) and (ii) *Cryptochaetum* (Agromyzidae) showing successive aggregation patterns of the flagellate *Polytoma*. (After Thorpe, 1930, 1932.)

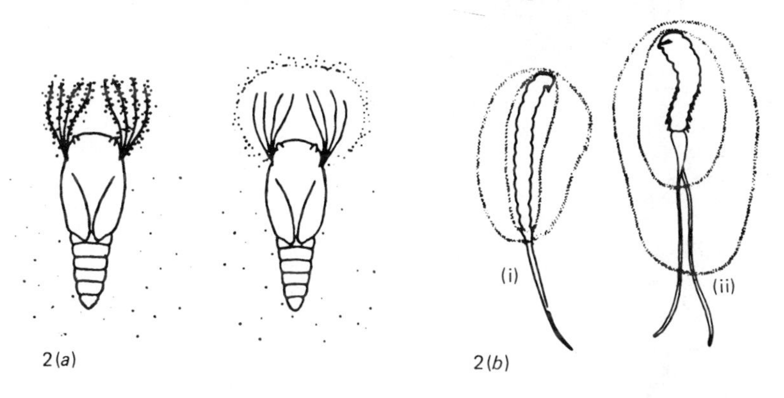

Fig. 3. Posterior extremity of larvae of *Culex pipiens* showing variation in size of anal papillae when reared in waters of different chlorinity. (After Wigglesworth, 1938.)

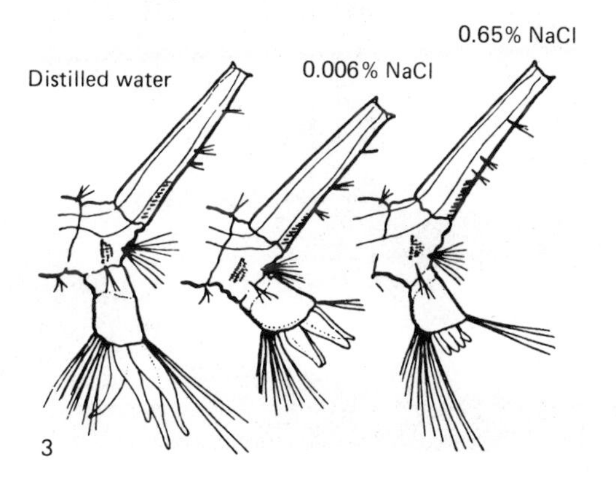

Fig. 4. Scanning electron micrograph of part of a filament of *Polypterus senegalensis* with secondary lamellae. Note the groups of cilia.

Fig. 5. Second gill arch of *Xyelacyba myersi*. Note the short length of the filaments.

Fig. 6. Perfusion-fixed specimen of *Lepidosiren paradoxa*. (*a*) Whole mount of single gill; note the absence of lamellae and presence of large capillary loops. (*b*) Semi-thin section of *Lepidosiren* gill showing capillaries of large diameter separated from water by thick water–blood barrier.

Fig. 7. Scanning electron micrograph of filament tip of *Huso huso* to show secondary lamellae.

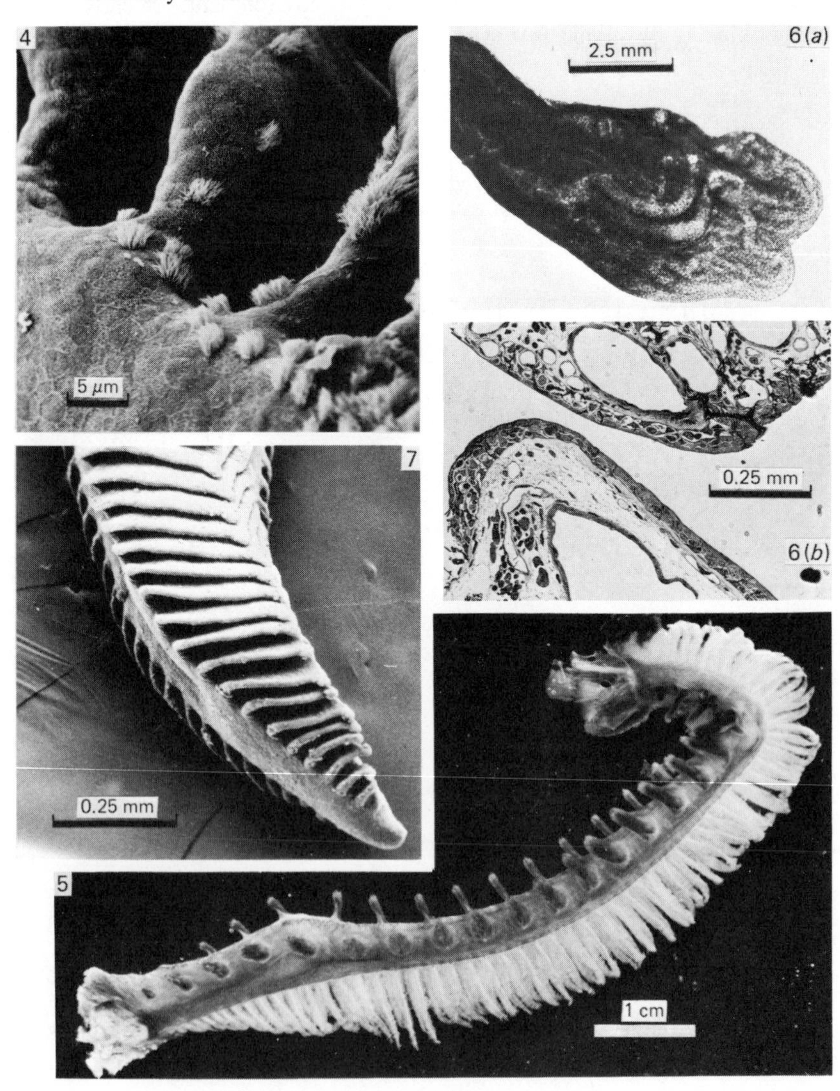

posterior end of mosquito larvae vary in size according to the salinity of the water in which they develop (Wigglesworth, 1938; Hopkins, 1967). A function in ionic regulation has been ascribed to them and perhaps for this reason they have often been referred to as papillae and not gills. It seems improbable that they are not concerned in gas exchange at least to some extent. From a respiratory point of view the degree of development of the external gills of salamanders is a well-known example (Dratisch, 1925; Bond, 1960), the gills being reduced in area and increased in thickness in larvae which live in oxygen-rich waters and vice versa in oxygen-depleted waters. From such an example it might be deduced that a gill arch such as that shown in Fig. 5 belongs to a fish also living in an oxygen-rich medium. In fact *Xyelacyba* inhabits deep waters (1400 m) of the Pacific and, as in many deep-sea fishes, the gill surfaces are poorly developed (Marshall, 1960; Hughes & Iwai, 1978).

These examples emphasise a most important feature of gills, namely that they form an interface for exchanges between the internal and external media and their structure and function respond to changes on both sides. Thus if the environmental oxygen supply is poor, a greater surface may tend to develop, and correspondingly, if the demand for oxygen on the inside is increased, there is also a tendency for the surface to become better adapted to provide that oxygen. Fig. 8 shows diagrammatically the large difference in oxygen tension at this interface in a fish gill and emphasises the importance not only of surface area but also thickness of the barrier to oxygen transfer. A number of cases exist even in recent zoological publications where surface area and degree of vascularisation are considered to be main factors determining the respiratory function and insufficient attention has been given to the importance of barrier thickness. In this

Fig. 8. Diagram of the water–blood barrier in a fish gill. Oxygen transfer across the different morphological layers of thickness t and total area A takes place due to the difference in partial pressure (ΔP_{O_2}). (After Hughes, 1970.)

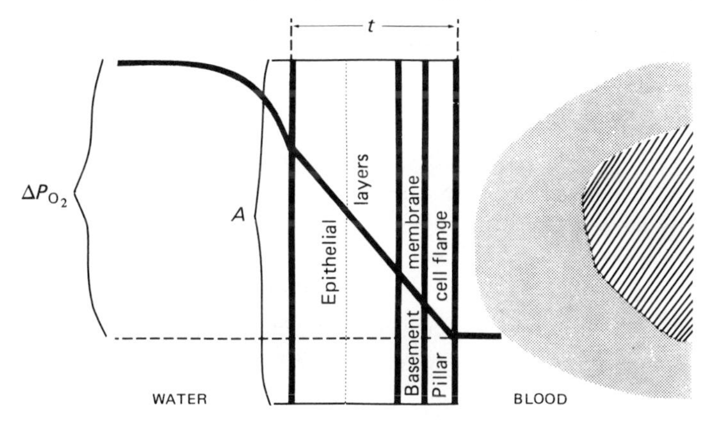

connection it should be remembered that although gills are generally outpushings of the body surface which increase the surface area for gas exchange, it is equally possible that localised regions of the body surface can become reduced in thickness and achieve the same functional result. Such examples occur in some invertebrates such as Terebellid worms (Wells & Dales, 1975; Weber, 1978). Perhaps such localised areas should also be called gills!

In other animals the presence of poorly-developed gills may be associated with the presence of alternative sites for gaseous exchange. Well-known examples are to be found among air-breathing teleosts such as mudskippers and *Anabas,* the climbing perch (Carter, 1957; Hughes & Munshi, 1968; Munshi, 1976), and other air-breathing fishes including the Dipnoi (Fig. 6). In spite of the reduced surfaces available on the gills of such fish they may, nevertheless, perform important functions, since the release of carbon dioxide from the blood tends to occur more into the water, via the gills and/or skin, whereas oxygen uptake is greater in the air-breathing organ (Hughes, 1966*b;* Singh, 1976). Furthermore, in some cases the area of the gills may be greater than that of the air-breathing organ, yet the latter may be more important in gas exchange because of the thinner barrier and hence greater diffusing capacity (Table 2).

Basic structure of gills

The gills of fishes (see Fig. 9 for their basic structure) are perhaps the most investigated aquatic respiratory organs in the animal kingdom and the following account is based upon them. (For more detailed reviews see Hughes & Morgan, 1973; Dunel & Laurent, 1980; and Hughes, 1980*b*.) Increases in gill surface area develop in relation to the water ventilation currents in such a way that close contact is ensured without an undue resistance being imposed to the flow of water (Hughes, 1966*a*). Correspondingly, the blood side of the exchanger comes into close contact with the surfaces across which the water flows. The flow of the two media is usually in opposite directions as this ensures more effective gas transfer (van Dam, 1938; Hughes & Shelton, 1962; Piiper & Baumgarten-Schumann, 1968; Hills & Hughes, 1970; Scheid & Piiper, 1976).

Fish gills consist of a large number of gill filaments arranged along the gill arches on each side of the pharynx. There are many variations in the gross morphology of the filaments and their attachment to the arch. In more primitive groups there is a well-developed septum which becomes reduced in teleosts. Coelacanths have an extension of the septum as in *Neoceratodus* (Fig. 10). Among teleosts there are variations in the degree of development of the septum (Dornescu & Miscalencu, 1968; Miscalencu, 1973; Hughes, 1980*b, c*) and a variety of secondary connections between neighbouring filaments occur in relation to ocean swimming, for example in tunas (Muir & Kendall, 1968), and air breathing, for example in *Amia* (Hughes & Morgan, 1973). The surfaces of the filaments are greatly enlarged by secondary lamellae (Fig. 7) which are the sites of gaseous exchange; their frequency along each side of the filaments ranges from 10 to 40/mm, being greater in more active species.

The essential structure of a secondary lamella is common to all fish and indeed to the gills of most aquatic animals. There are two sheets of epithelia separated by spaces through which the blood circulates. The separation between the epi-

Fig. 9. Diagrams to illustrate the basic structure of a fish gill. (*a*) Part of a single gill arch showing the path of water between gill filaments and secondary lamellae. Note the counter flow with blood. (*b*) Profile of a gill sieve between two filaments with secondary lamellae. (*c*) Diagram of secondary lamellae showing paths for blood flow between pillar cells from the afferent (a) to the efferent (e) filament arteries. The distribution of chloride (c) and mucous (M and heavy stipple) cells is also indicated. (*d*) Diagrammatic section of a secondary lamella based on electron micrographs. (After Bettex-Galland & Hughes, 1973 and Hughes, 1980*b*.)

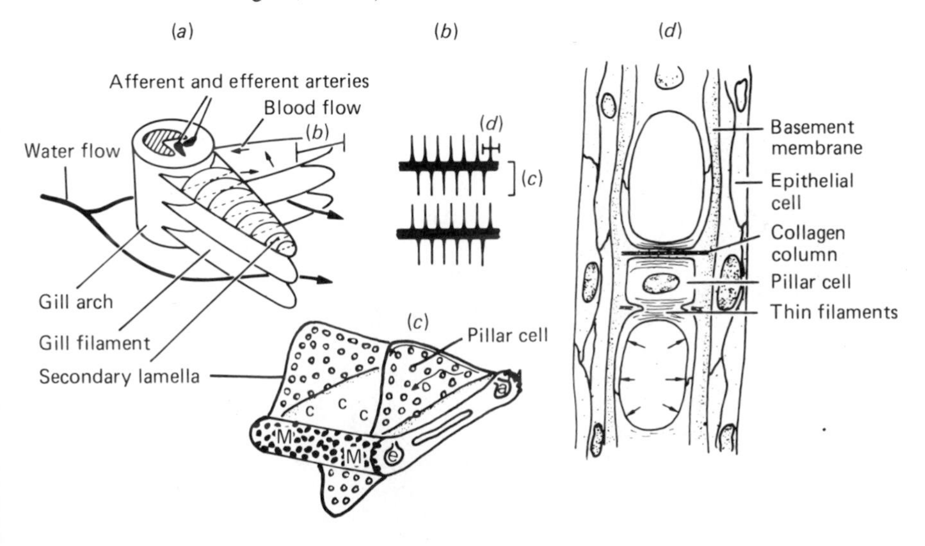

Fig. 10. Diagrams to show differences in arrangement of gill filaments among the main groups of fishes. (From Hughes, 1980*c*.)

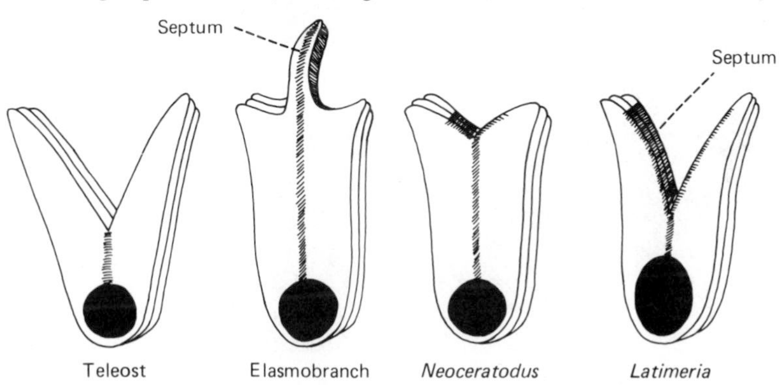

thelial sheets is maintained by a series of pillars which in the fish gill are formed by the pillar cells. Between the pillar cells and the epithelial layers the basement membranes form the structural skeleton of these lamellae. The basement membrane has several layers, the innermost of which is collagenous, and these layers are connected by groups of columns each of which is enfolded by the cytoplasm of a pillar cell body (Figs. 8, 9, 11). The pillar cell is a highly characteristic

Fig. 11. Section of a secondary lamella of *Latimeria chalumnae* showing pillar cell with single column of large diameter. (After Hughes, 1980*c*.)

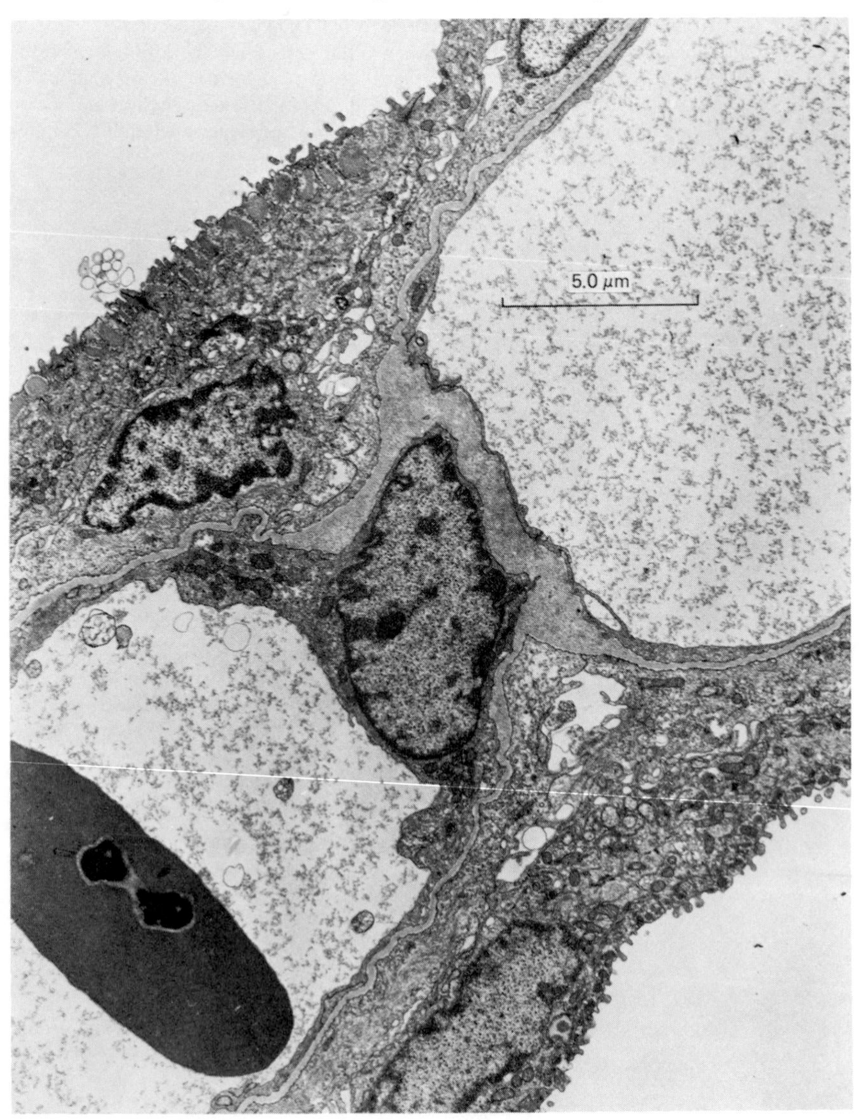

structure of fish gills and even in such a special fish as *Latimeria,* the same essential organisation is present (Fig. 11). The presence of pillar cells has made it possible to identify structures which have been derived from gills to form complex air-breathing organs (Munshi, 1968; Hughes & Munshi, 1973). From a functional point of view the presence of many fine filaments within the pillar cells has given rise to the view that they might have a contractile function and so regulate blood flow within secondary lamellae (Hughes & Grimstone, 1965). Evidence for an actomyosin–myosin mechanism has been obtained (Bettex-Galland & Hughes, 1973) and has recently received confirmation by Smith & Chamley-Campbell (1981) using an immunofluorescence technique. In fact Biétrix (1895) was the first to name these cells and in his thesis he also suggested a contractile function. His views on the nature of the columns did not accord with their containing collagen and he believed that there was only one layer of epithelial cells outside the basement membrane.

Much information is now available about the detailed structure of secondary lamellae and the presence of a lymphoid space between the two epithelial layers has been established (Hughes & Wright, 1970). Its precise function has still to be elucidated but it certainly contains a variety of white blood cells, some of which might perform functions analogous to alveolar macrophages (Hughes & Gray, 1972), and a circulation of its fluid which may be important in the protective and osmoregulatory function of gills has been suggested (Hughes, 1978).

It is evident that the extent of the gill surface is related to the number of filaments (primary lamellae) as well as the frequency of the secondary lamellae along them. Among fishes, filament number does not increase as markedly as the number of secondary lamellae during the growth of a particular animal. As will become apparent later in this chapter, the gills of crustaceans also show less dependence upon an increase in the number of primary lamellae (secondary lamellae being absent), but the area of these lamellae increases most significantly.

The secondary lamellae of different species of fish vary very much in their detailed morphology, most obviously in shape, and heterogeneity is also found in different parts of the gill system of a particular species (Hughes, 1973). Those shown in Fig. 7 have a triangular shape at the tips of the filaments whereas nearer their base they are more rectangular. In some species (e.g. *Barbus Sophor,* Hughes & Mittal, 1980) the surface of individual lamellae is further increased by tertiary lamellae but their importance for gas exchange is limited. The filaments of this species also emphasise another general feature in that most of the pillar-cell system (i.e. pillar cells plus associated blood channels) is deeply embedded within the main tissue of the filament and is greatly removed from the water surfaces. Thus from a circulatory point of view each secondary lamella is by no means uniform in structure as the channels through which blood circulates are separated from the water by varying distances. The most exposed blood is probably that which circulates through the marginal channels along the free edge of each secondary lamella. This channel also differs from the others in that its outer wall is lined by true endothelial cells whereas the other channels are bordered by pillar-cell bodies and their flanges (Fig. 9*d*). It would appear that pillar cells are not homologous with endothelial cells but are derived directly from mesenchyme (Tovell, Morgan & Hughes, 1970; Morgan, 1974). Contraction of the pillar cells would tend to restrict blood flow to the marginal and proximal channels of each secondary lamella. Thus recruitment of blood channels can occur within individual lamellae as well as whole lamellae being unperfused (Rankin & Metz, 1971; Hughes, 1972*a*; Booth, 1978).

The surface of secondary lamellae is not smooth but is covered by many microvilli and microridges (Hughes & Wright, 1970; Hughes, 1979*a*) which would appear to increase still further the exchange surface (Lewis & Potter, 1976). However, the extent to which this is functional depends upon the condition of the microridged surface during normal life. The space between the ridges appears to be filled with mucus and the anchoring of such a layer may be one of the main functions of the microridges. If that is so, then the effective outer surface of the secondary lamella must be considered as a fairly smooth one at the level of the tips of the microvilli or ridges (Hughes, 1979*a*).

The gills of invertebrates illustrate the same general principles as those of fish (Yonge, 1947; Copeland & Fitzjarrell, 1968; Fisher, 1972; Mill, 1972; Taylor & Greenaway, 1979). Increase in total gill surface is achieved by branching of the main gill axis, for example in cephalopods where the branching may be at least tertiary (Fig. 12). Sections across the exchange surfaces also reveal a similar organisation into two sheets of epithelia joined together periodically by trabeculae or trabecular cells which are analogous to pillar cells (Nakao, 1976), which strictly speaking are only found in fish gills. The measurement of surface areas of molluscan gills is much more difficult than for fish or crustaceans, but in all groups a simple approach is to weigh the gill and this may give a relation to body weight (Fig. 13) which is similar to that for the gill surface. The slope of this log versus log regression is 0.85, which is close to that first established for tunas (Muir & Hughes, 1969). However, for fish gills the relation between gill weight and body weight is not always so similar (Hughes, 1980*c*).

The compactness of gill systems is well known and can be observed in crabs if a small plastic window is inserted into the branchiostegite (Hughes, Knights & Scammell, 1969). A similar method was used by Bijtel (1949) to observe fish gills during the normal ventilatory cycle. She established that the gill sieve is by no means a rigid and passive system, yet the filament tips of adjacent arches remain in close contact. Crab gills are less mobile so that samples of water can be taken from between individual lamellae and the counter-flow exchange between water and blood can be thus confirmed. As mentioned earlier, the main axis of crab gills is subdivided into many primary lamellae which lie very close to one another (Fig. 14*a*). Scanning electron micrographs at the cut edge of these structures show that there are trabecular processes connecting the two surfaces which in this case are formed of epithelial cells supported by chitinous rods, and the outermost surface appears to be smooth (Fig. 14*b*) although at higher magnification it has a highly organised fine structure (Filshie & Smith, 1980). Blood flows between the trabeculae and there is also a well-defined marginal channel.

Morphometry of gill surfaces in relation to body size

In relation to both their respiratory and osmoregulatory function it is often important to have a quantitative estimate of the exchange surface area and thickness of the barrier separating the blood and water. Relatively few detailed studies have been made among animals but the studies of Dr I. E. Gray (1954, 1957) of Durham, North Carolina are notable in this context. Among a series of American crabs Gray (1957) established that, largely due to an increase in lamellar area, the total surface increased during the growth of individual species. He concluded that the gill areas of more active crabs are greater than those of sluggish species. Analysis of some of the data kindly made available by Dr Gray has been carried out using logarithmic transformation, and the following relation has been established.

$\log_{10}$ gill area = $\log_{10}$ area of 1 g animal (a) + $b \times \log_{10}$ body mass (W)

or area = aW^b where b is the slope of the regression analysis

Fig. 12. One of the main branches of a single gill of *Alloteuthis* to show branching arrangement and absence of typical lamellae.

Fig. 13. Bi-logarithmic plot to show relation between weight of gill and body weight (W) of 30 specimens of *Aplysia fasciata* collected at Naples in 1973. Brackets show 95% confidence limits of regression line at 10, 100 and 1000 g.

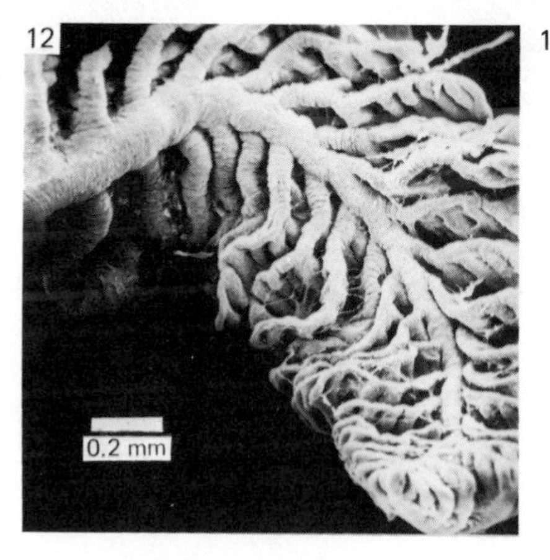

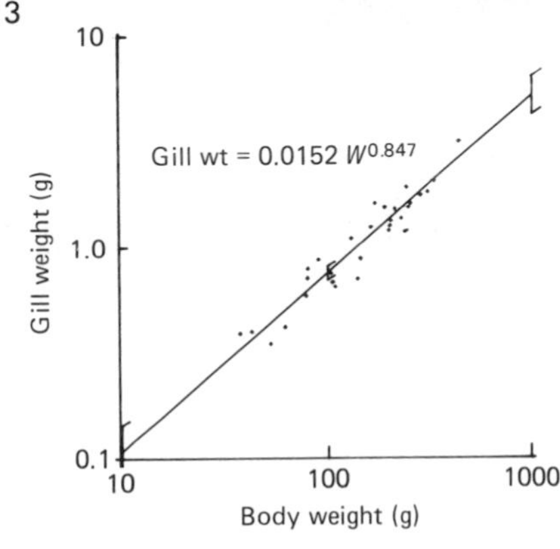

Fig. 14. Scanning electron micrographs of *Carcinus maenas*. (*a*) To show arrangement of lamellae on a single gill axis. (*b*) Higher magnification of lamellae cut across to reveal presence of trabeculae (arrows).

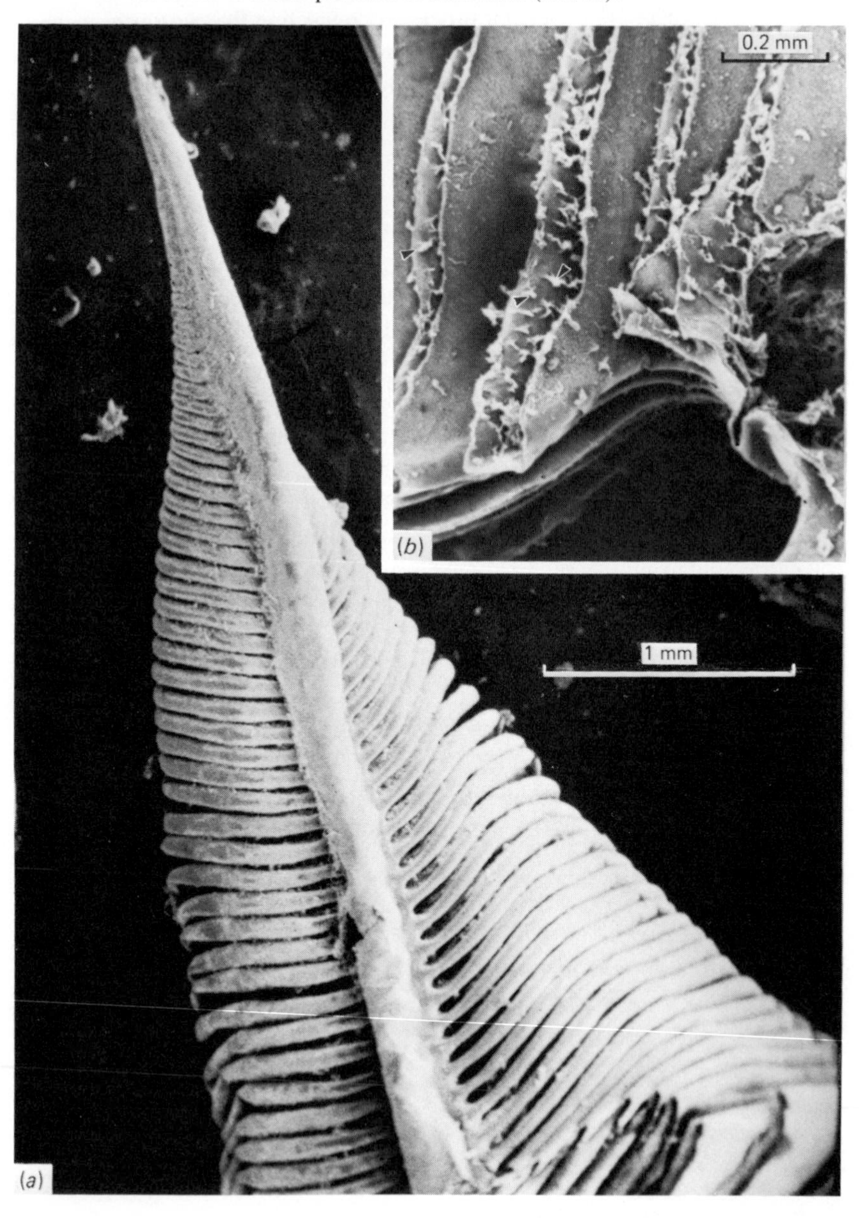

As gill area = total number (N) of lamellae × bilateral area of an average lamella ($b\ell$)

then, Area $= a'a''W^{(b'+b'')}$, where a', b' and a'', b'' are the a and b values obtained by logarithmic transformation of measurements of N and $b\ell$ respectively.

The results of such an analysis for eight of the species he used is given in Table 1 where it can be seen that the slope (b) of the regression line varies between 0.5 and 1.0. It is apparent that this relation for total gill surface is made up of the sum of the relation for total number of lamellae and the area of an average lamella. The analysis shows that the latter is the more important component. With the large amount of data available, it was of interest to take mean values for area, N and $b\ell$ for each of the eight species and to use these mean values for separate regression analysis. Such an interspecific plot for gill area has a slope of about one. This value can only be taken as typical for one of the eight species and illustrates the misleading nature of interspecific plots to represent intraspecific relations. In fact, this example is very similar to one discussed from a purely theoretical point of view (Hughes, 1980*d*). One disadvantage of the intraspecific plots is that inevitably they cover a smaller range of body mass, and hence the confidence limits of the regression lines tend to be wider.

Similar results to those obtained with the American species have also been

Table 1. *A. Values for slope* (b) *of the regression line obtained by plotting values for gill area and component measurements for eight species from Gray (1954) on bi-logarithmic coordinates. B. Results of a similar analysis based on the mean values for each of the eight species* (*see text*)

Species	No. of animals N	Mean body weight (g)	Gill area	No. of platelets	Average platelet[a] area
A. *Ocypode albicans*	24	41.35	0.795	0.115	0.680
Uca minax	33	6.56	0.543	0.055	0.480
Sesarma cinerea	13	1.48	0.731	0.258	0.474
Panopeus herbstii	36	14.99	1.003	0.098	0.906
Menippe mercernaria	60	149.32	0.823	0.194	0.631
Libinia emarginata	26	127.14	0.775	0.157	0.619
L. dubia	11	109.67	0.670	0.083	0.587
Callinectes sapidus	39	113.75	0.935	0.156	0.779
B. Interspecific means	8	70.53	1.022	0.328	0.688

[a] It should be noted that Gray (1954) used the term 'platelet' to denote the pair of lamellae at the same level of a gill axis. Thus 1 platelet = 2 lamellae.

found by analysis of some measurements on British decapod crustaceans (Hughes, 1982). Regression lines for two of these are plotted in Fig. 15 which also includes the relations for six fish species and for the mammalian lung. In the latter case a subdivision of the species into 'active' and 'inactive' groups has been made (Weibel, 1973) but both of these interspecific plots have slopes of 0.9–1.0. None of the other species have respiratory organs with such large surface areas as the mammalian lung, although those of the most active fishes (tuna and *Coryphaena*) and crustaceans (*Callinectes*) are close to the mammalian values for similar body weights. Few data are available for molluscan respiratory organs (see Ghiretti,

Fig. 15. Bi-logarithmic plot of surface areas of respiratory organs of a range of vertebrates and invertebrates plotted against body weight. Full lines indicate crustaceans, dashed lines indicate fishes and lines for 'active' and 'inactive' species of mammals are also given. Measurements for single specimens of *Buccinum* and *Nautilus* (asterisks) and *Latimeria* (filled circle) are also plotted.

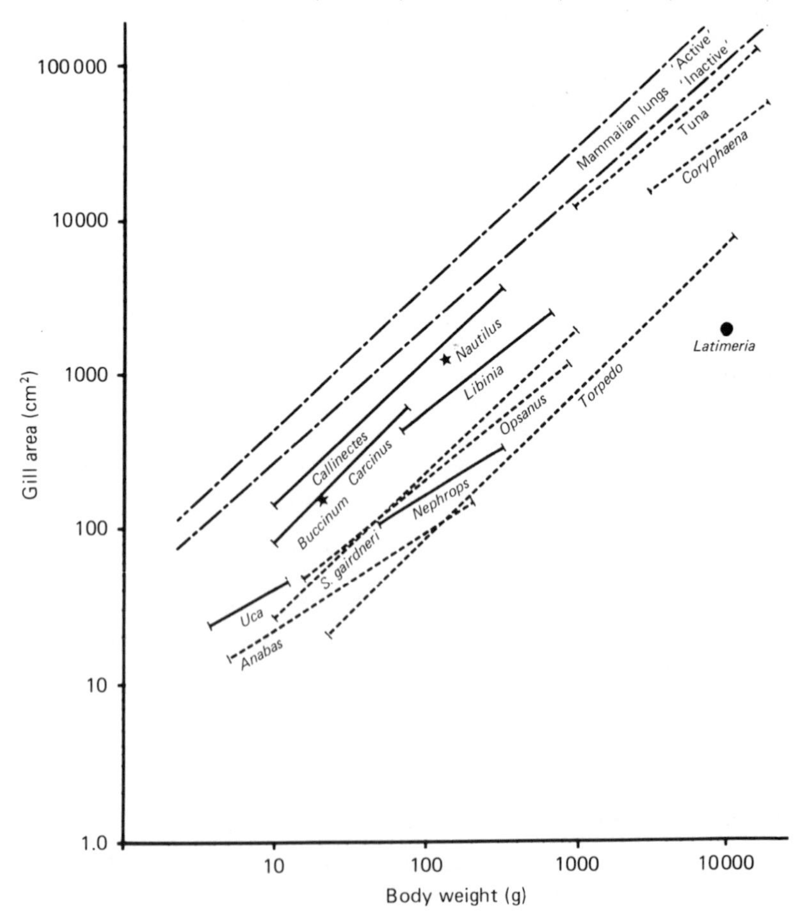

1966), but the figure for *Nautilus* is relatively large, being about 10cm^2 g^{-1}. In all groups the less active species tend to have the smaller areas and among fishes the lowest are found in some deep-water marine species including the coelacanth *Latimeria* (Fig. 15).

As has been indicated earlier, it is important to take into consideration the thickness of the barrier between the water and blood and this can be summarised by giving values for the diffusing capacity, which can be estimated from the morphometric measurements (Fig. 8) as follows (for details see Hughes, 1972*a*):

Rate of oxygen transfer, $\dot{V}_{O_2} = \overline{\Delta} P_{O_2} KA/t$, where K is the Krogh permeation coefficient and $\overline{\Delta P}_{O_2}$ is the mean P_{O_2} difference between blood and water

Hence, $\dot{V}_{O_2}/\overline{\Delta} P_{O_2} = K\,A/t =$ Diffusing capacity

Estimates for the morphometric diffusing capacity of a number of gills are included in Table 2; several exceptions are given to the general finding (Hughes, 1970) that fish gills with larger surface areas tend to have thinner barriers and consequently much greater diffusing capacities. Unfortunately, there are many possible sources of error in these calculations and it is not always certain that the measurements between different species are comparable because they have been obtained by different methods. In general the area measurements are the more reliable in spite of differences in sampling methods, as there has been little consistency in the way of expressing the thickness of the water–blood barrier. Furthermore, some authors have emphasised the fact that only a part of the total gill surface area overlies the blood capillaries and consequently have used the term 'respiratory area' for that part of the total area which is fully vascularised. At most respiratory surfaces, however, it is apparent that gas exchange takes place across the whole surface but varies in amount according to the diffusion distances between the surface and the nearest blood-pigment molecule. Probably the most reliable method is to use the total surface area together with the harmonic mean of the diffusing distances from the outer epithelial surface to the nearest blood channel. These measurements are made on randomly-sectioned material by using a rectilinear grid (Hughes, 1972*a;* Hughes & Weibel, 1976). Such a method is generally applied to mammalian lungs but has been applied in one or two cases to gill surfaces. It is more common for the thickness to be an expression of the mean or perhaps minimum thickness across the regions overlying the blood channels. If such mean values are taken for these limited diffusion channels, then, correspondingly, the area measurement must be reduced. The overall values obtained for diffusing capacity using these two methods are close to one another (Fig. 16) provided that the blood channels are relatively large (*c*) as in most gills. For more accurate measurements in other gills it is necessary to use a mean value of the surface area of the blood channels and area of the outer part of the lamellar

Table 2. *Oxygen diffusing capacity of the tissue barrier* (D_t) *and component measurements for respiratory surfaces of seven species of fish and two invertebrates*

Species and organ	Body wt (g)	Thickness[a] of tissue barrier (μm)	Surface area g^{-1} (mm^2 g^{-1})	Diffusing[b] capacity (ml min^{-1} $Torr^{-1}$ kg^{-1})	Reference
Anabas testudineus					
All gills	100	10.00	47.2	0.0071	Hughes, Dube & Munshi (1973)
Labyrinthine organs	100	0.21	32.0	0.2286	
Heteropneustes fossilis					
All gills	100	3.58	57.7	0.0242	Hughes *et al.* (1974)
Air sac	100	1.60	30.7	0.0288	
Skin	100	98.00	200.0	0.0031	
Salmo gairdneri					
All gills	35	4.30 (τ)	260.0	0.1180	Hughes & Perry (1976)
Tinca tinca					
All gills	141	2.47 (τ)	228.0	0.1493	Hughes (1972*a*)
Tuna					
All gills	100	0.50	2000.0	6.0000	Hughes & Gray (1972)
Opsanus tau					
All gills	100	5.00	210.0	0.0630	
Latimeria chalumnae					
All gills	10^3	5.00	18.9	0.0057	Hughes (1972*b*)
Carcinus maenas					
All gills	100	5.00 (2.00 chitin)	744.0	0.0540	Scammell & Hughes (1981)
Nautilus macromphalus					
All gills	135	10.00	930.0[c]	0.1395	—

[a] Thickness given is arithmetic mean of thin parts of barrier, except in two cases where τ = harmonic mean.
[b] Value for permeation coefficient (K) used has been 1.5×10^{-6} ml O_2 mm^{-2} μm^{-1} $Torr^{-1}$ except for *S. gairdneri* (see Hughes & Perry, 1976) and for chitinous part of barrier in *Carcinus maenas* ($K = 1.71 \times 10^{-7}$ ml O_2 mm^{-2} μm^{-1} $Torr^{-1}$).
[c] Pelseneer, P. (1935).

surface (Hughes & Perry, 1976). Distances should also be measured to the nearest red blood cell when these are present.

It would greatly improve the ease of comparison of different gills as gas exchange structures if a consistent method were adopted. In the present situation it is important for each worker to make it quite clear which area and thickness measurements have been used in any calculations.

Gas exchange

Perhaps the ultimate aim of many respiratory physiologists when they investigate gills is to be able to give a quantitative expression for the distribution of the partial pressure of oxygen (P_{O_2}) and the partial pressure of carbon dioxide (P_{CO_2}) at all stages of the exchange processes in single lamellae. Unfortunately, it is very difficult to make any direct measurements especially if the secondary lamellae retain their normal morphological and physical relations. As indicated

Fig. 16. Diagrams to illustrate two methods for measuring the thickness and surface area of respiratory structures for use in estimating diffusing capacity. The upper surface has an area A and lines indicate distances for measurement of harmonic means. Lower surfaces show areas (A') over blood channels and lines for measurements used to obtain diffusion distances. Blood channels increase in relative size from (a) to (c).

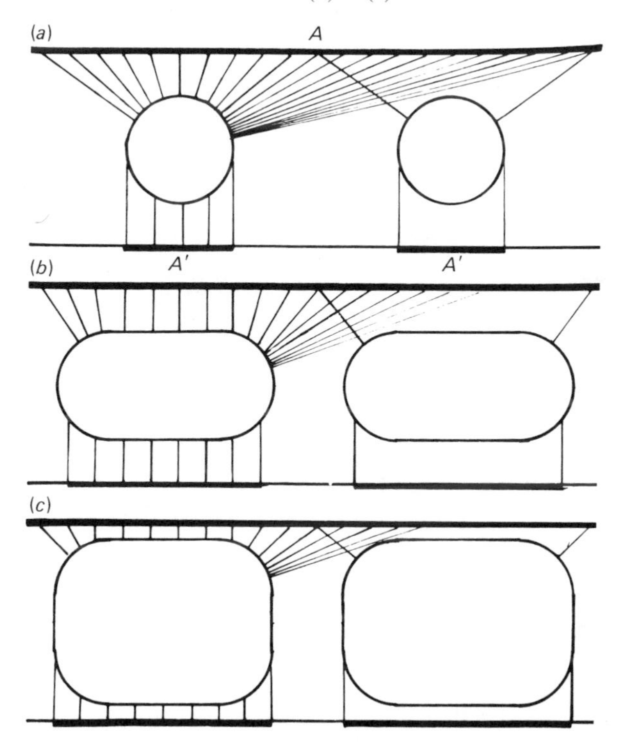

earlier the stiffer gills of crabs have enabled water samples to be taken from different parts of the interlamellar spaces, but as yet the corresponding P_{O_2}s in the blood are only known at the two ends of the lamella (Fig. 17). The exposed nature of the exchange surfaces of fish gills, however, makes them more acces-

Fig. 17. Diagrams indicating P_{O_2} of water samples taken from different interlamellar regions of *Carcinus*. In addition two points show the P_{O_2} of the afferent and efferent blood. (After Hughes, Knights & Scammell, 1969; Scammell, 1971.)

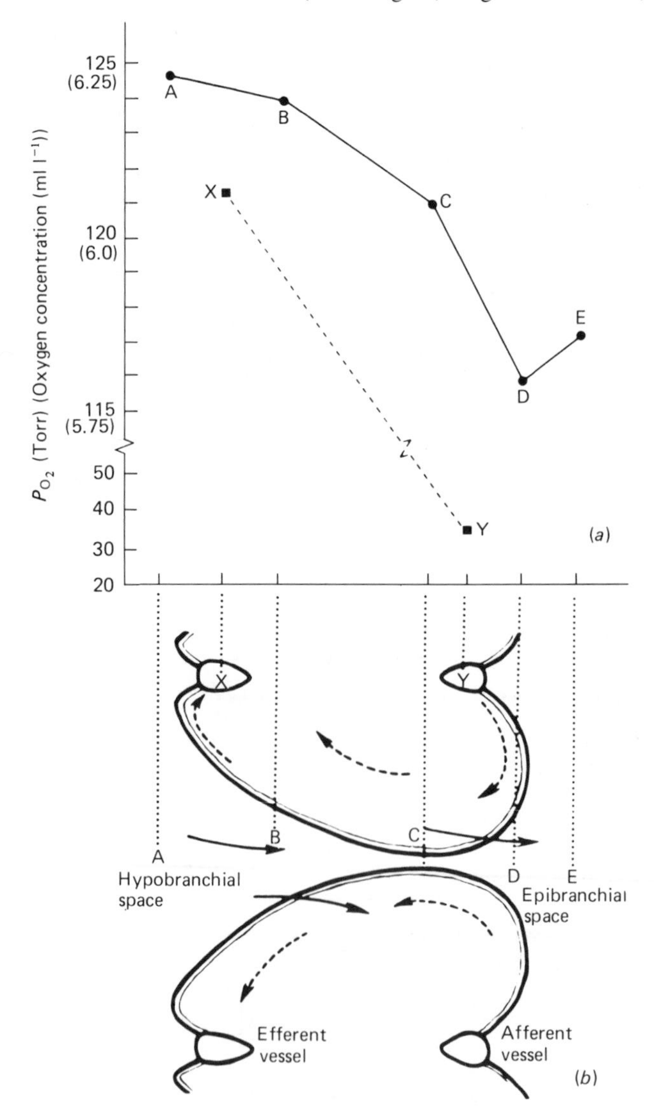

sible. It has been possible, for example, to isolate single secondary lamellae and to obtain data on the time course of oxygenation and deoxygenation of single red blood cells (Hughes & Koyama, 1974). Measurements have also been made of the blood flow velocity in different parts of the filament circulation (Hughes *et al.*, 1981). These indicate that the blood remains in contact with the water for about 0.5 s and this is the time required for its complete oxygenation.

Another approach to this problem has been to construct models of the gill system. In some models, direct measurement of blood flow or its equivalent can be made (Olson, 1979), and in other cases electrical analogue models have proved of value. For example, circulation in a single gill filament of a fish may be simulated by a network of resistances which indicate that blood flow across lamellae at the tips of filaments tends to be less than in lamellae at the base of filaments (Hughes, 1980*b*). Estimates of current flow in the resistances corresponding to those of the secondary lamellae in more complex networks can be made in computer simulations, insertion of different values for parameters such as the compliance of the secondary lamellae indicate that a reduction in lamellar compliance will favour an increase in perfusion of the more distal lamellae (G. M. Hughes & A. Belaud, unpublished observations).

In such modelling, the more detailed information regarding the complexity of blood pathways within fish gills (see Vogel, Vogel & Kremers, 1973; Vogel, 1978; Laurent, 1981) makes the situation far more complex than was originally supposed. Furthermore, it is probable that the 'blood' passing along the different pathways may not have the same haematocrit value (Hughes, 1979*b*). This can give rise to differences in resistance to blood flow even though the nucleated red cells of fish blood have been shown to be very deformable (Hughes, Kikuchi & Watari, 1982). The presence of shunt pathways within the gill circulation was first given prominence by Steen & Kruysse (1964) for the eel, and although the presence of direct shunts between afferent and efferent filament arteries is very improbable, there does nevertheless seem to be good evidence that, at least in some species, a certain proportion of the cardiac output does not come into direct contact with the gas exchange surfaces and is returned directly to the heart without passing into the dorsal aorta (Hughes, Peyraud, Peyraud-Waitzenegger & Soulier, 1981).

The whole subject of fish gill structure and function is, therefore, at a very interesting stage, but what is quite certain is that the classical description of the gill circulation and function needs to be modified in quite a radical way although its precise nature in different species is not certain at the moment. It is also evident that the basic structure of nearly all gills, that is, two permeable sheets separated by regularly placed pillars, is highly adapted, and has already been copied by man in some important medical and engineering applications.

References

Baskin, D. G. & Detmers, P. A. (1976). Electron microscopic study on the gill bars of *Amphioxus* (*Branchiostoma californiense*) with special reference to neurociliary control. *Cell and Tissue Research,* **166,** 167–78.

Bettex-Galland, M. & Hughes, G. M. (1973). Contractile filamentous material in the pillar cells of fish gills. *Journal of Cell Science,* **13,** 359–66.

Biétrix, E. (1895). Etude de quelques faits relatifs à la morphologie générale du système circulatoire a propos du réseau branchial des poissons. Thèse, 56 pp. Faculté de Médecine, Paris.

Bijtel, J. H. (1949). The structure and the mechanism of movement of the gill filaments in Teleostei. *Archives Néerlandaises des Sciences Exactes et Naturelles,* **8,** 267–88.

Bond, A. N. (1960). An analysis of the response of salamander gills to changes in the oxygen concentration of the medium. *Developmental Biology,* **2,** 1–20.

Booth, J. H. (1978). The distribution of blood flow in the gills of fish: application of a new technique to rainbow trout (*Salmo gairdneri*). *Journal of Experimental Biology,* **73,** 119–29.

Boyle, R. (1660). *New Experiments Physico-Mechanicall, touching the spring of the Air, and its Effects* (*Made for the most part, in a New Pneumatical Engine*), 207 pp. Oxford: Robinson.

Carter, G. S. (1957). Air Breathing. In *Physiology of Fishes,* vol. 1, ed. M. E. Brown, pp. 65–79. New York: Academic Press.

Copeland, D. E. & Fitzjarrell, A. T. (1968). The salt absorbing cells in the gills of the blue crab (*Callinectes sapidus* Rathbun) with notes on modified mitochondria. *Zeitschrift für Zellforschung und Mikroskopische Anatomie,* **92,** 1–22.

Crisp, D. J. (1964). Plastron breathing. In *Recent Progress in Surface Science,* vol. 2, ed. J. F. Danielli, K. G. A. Pankhurst & A. C. Riddiford, pp. 377–425. New York: Academic Press.

Dam, L. van (1938). On the utilisation of oxygen and regulation of breathing in some aquatic animals. Dissertation. Groningen.

Dornescu, G. T. & Miscalencu, D. (1968). Cele trei tipuri de branhii ale teleosteenilor. *Analele Universitatii Bucuresti Seria Stiintele Naturii Biologie Anul,* **XVII,** 11–20.

Dratisch, L. (1925). Über das Leben der Salamandra-Larven bei hohem und niedrigem Sauerstoffpartialdrück. *Zeitschrift für Vergleichende Physiologie,* **2,** 632–57.

Dunel, S. & Laurent, P. (1980). Functional organisation of the gill vasculature in different classes of fishes. In *Epithelial Transport in the Lower Vertebrates,* ed. B. Lahlou, pp. 37–58. Cambridge University Press.

Ege, R. (1915). On the respiratory function of the air stores carried by some aquatic insects (Corixidae, Dytiscidae & *Notonecta*). *Zeitschrift für Allgemeine Physiologie,* **17,** 81–124.

Filshie, B. K. & Smith, D. S. (1980). A proposed solution to a fine-structural puzzle. The organisation of gill cuticle in a crayfish (*Panulirus*). *Tissue and Cell,* **12,** 209–26.

Fisher, J. M. (1972). Fine structural observations on the gill filaments of the freshwater crayfish, *Astacus pallipes* Lereboullet. *Tissue and Cell,* **4,** 287–99.

Fox, H. M. (1921). Methods of studying the respiratory exchange in small aquatic organisms, with particular reference to the use of flagellates as an indicator for oxygen consumption. *Journal of General Physiology,* **3,** 565–73.

Ghiretti, F. (1966). Respiration. In *Physiology of Mollusca,* ed. K. M. Wilbur & C. M. Yonge, vol. 2, pp. 175–208. London & New York: Academic Press.

Gray, I. E. (1954). Comparative study of the gill area of marine fishes. *Biological Bulletin of the Marine Biological Laboratory, Woods Hole,* **107,** 219–25.

Gray, I. E. (1957). A comparative study of the gill area of crabs. *Biological Bulletin of the Marine Biological Laboratory, Woods Hole,* **112,** 34–42.

Hills, B. A. & Hughes, G. M. (1970). A dimensional analysis of oxygen transfer in the fish gill. *Respiration Physiology,* **9,** 126–40.

Hinton, H. E. (1966). Plastron respiration in marine insects. *Nature, London,* **209,** 220–1.

Hopkins, C. R. (1967). The fine structural changes observed in the rectal papillae of the mosquito *Aedes aegypti,* L. and their relation to the epithelial transport of water and inorganic ions. *Journal of the Royal Microscopical Society,* **86,** 235–52.

Horman, W. (1519). *Vulgaria,* 277b.

Hughes, G. M. (1966*a*). The dimensions of fish gills in relation to their function. *Journal of Experimental Biology,* **45,** 177–95.

Hughes, G. M. (1966*b*). Evolution between air and water. In Ciba Foundation Symposium on *Development of the Lung,* ed. A. V. S. de Reuck & R. Porter, pp. 64–80. London: Churchill.

Hughes, G. M. (1970). Morphological measurements on the gills of fishes in relation to their respiratory function. *Folia Morphologica* (Prague), **18,** 78–95.

Hughes, G. M. (1972*a*). Morphometrics of fish gills. *Respiration Physiology,* **14,** 1–25.

Hughes, G. M. (1972*b*). Gills of a living coelacanth *Latimeria chalumnae. Experientia,* **28,** 1301–2.

Hughes, G. M. (1973). Comparative vertebrate ventilation and heterogeneity. In *Comparative Physiology,* ed. L. Bolis, K. Schmidt-Nielsen & S. H. P. Maddrell, pp. 187–220. Amsterdam: North-Holland Publishing Company.

Hughes, G. M. (1978). Morphology and morphometrics of fish gills. *Atti della Societa Peloritana Science Fisiche Matematicha e Naturali Messina,* **24,** 333–5.

Hughes, G. M. (1979*a*). Scanning electron microscopy of the respiratory surface of trout gills. *Journal of Zoology,* **188,** 443–53.

Hughes, G. M. (1979*b*). The path of blood flow through the gills of fishes – some morphometric observations. *Acta morphologica Sofia,* **2,** 52–8.

Hughes, G. M. (1980*a*). Structure of fish respiratory surfaces. *International Congress of Anatomy, Mexico City,* Abs 36.

Hughes, G. M. (1980*b*). Functional morphology of fish gills. In *Epithelial Transport in the Lower Vertebrates,* ed. B. Lahlou, pp. 15–36. Cambridge University Press.

Hughes, G. M. (1980*c*). Ultrastructure and morphometry of the gills of *Latimeria chalumnae,* and a comparison with the gills of associated fishes. *Proceedings of the Royal Society of London, Series B,* **208,** 309–28.

Hughes, G. M. (1980*d*). Morphometry of fish gas exchange organs in relation to their respiratory function. In *Environmental Physiology of Fishes,* ed. M. A. Ali, pp. 33–56. New York: Plenum.

Hughes, G. M. (1982). Allometry of gill dimensions in some British and American decapod Crustacea. In preparation.
Hughes, G. M., Dube, S. C. & Munshi, J. S. D. (1973). Surface area of the respiratory organs of the climbing perch, *Anabas testudineus* (Pisces-Anabantidae). *Journal of Zoology,* **170,** 227–43.
Hughes, G. M. & Gray, I. E. (1972). Dimensions and ultrastructure of toadfish gills. *Biological Bulletin of the Marine Biology Laboratory, Woods Hole,* **143,** 150–61.
Hughes, G. M. & Grimstone, A. V. (1965). The fine structure of the secondary lamellae of the gills of *Gadus pollachius. Quarterly Journal of Microscopical Science,* **106,** 343–53.
Hughes, G. M., Horimoto, M., Kikuchi, Y., Kakiuchi, Y. & Koyama, T. (1981). Blood flow velocity in microvessels of the gill filaments of goldfish (*Carassius auratus* L.). *Journal of Experimental Biology,* **90,** 327–31.
Hughes, G. M. & Iwai, T. (1978). A morphometric study of the gills in some Pacific deep-sea fishes. *Journal of Zoology,* **184,** 155–70.
Hughes, G. M., Kikuchi, Y. & Watari, H. (1982). A study of the deformability of red blood cells of a teleost fish, the yellow-tail (*Seriola quinqueradiata*), and a comparison with human erythrocytes. *Journal of Experimental Biology,* **96,** 209–20.
Hughes, G. M., Knights, B. & Scammell, C. A. (1969). The distribution of P_{O_2} and hydrostatic pressure changes within the branchial chambers in relation to gill ventilation of the shore crab *Carcinus maenas* L. *Journal of Experimental Biology,* **51,** 203–20.
Hughes, G. M. & Koyama, T. (1974). Gas exchange of single red blood cells within secondary lamellae of fish gills. *Journal of Physiology,* **246,** 82–83P.
Hughes, G. M. & Kylstra, J. L. (1964). Breathing in water. *New Scientist,* **24,** 566–9.
Hughes, G. M. & Mittal, A. K. (1980). Structure of the gills of *Barbus sophor* (Ham), a cyprinid with tertiary lamellae. *Journal of Fish Biology,* **16,** 461–7.
Hughes, G. M. & Morgan, M. (1973). The structure of fish gills in relation to their respiratory function. *Biological Reviews,* **48,** 419–75.
Hughes, G. M. & Munshi, J. S. D. (1968). Fine structure of the respiratory surfaces of an air-breathing fish, the climbing perch *Anabas testudineus* (Bloch). *Nature, London,* **219,** 1382–4.
Hughes, G. M. & Munshi, J. S. D. (1973). Nature of the air-breathing organs of the Indian fishes, *Channa, Amphipnous, Clarias* and *Saccobranchus* as shown by electron microscopy. *Journal of Zoology,* **170,** 245–70.
Hughes, G. M. & Perry, S. F. (1976). Morphometric study of trout gills: a light microscopic method suitable for the evaluation of pollutant action. *Journal of Experimental Biology,* **64,** 447–60.
Hughes, G. M., Peyraud, C., Peyraud-Waitzenegger, M. & Soulier, P. (1981). Proportion of cardiac output concerned with gas exchange in gills of the eel (*A. anguilla*). *Journal of Physiology,* **310,** 61–2P.
Hughes, G. M. & Shelton, G. (1962). Respiratory mechanisms and their nervous control in fish. In *Advances in Comparative Physiology and Biochemistry,* ed. O. Lowenstein, vol. 1, pp. 275–364. London: Academic Press.
Hughes, G. M., Singh, B. R., Guha, G., Dube, S. C. & Munshi, J. S. D. (1974). Respiratory surface areas of an air-breathing siluroid fish, *Saccobranchus (Heteropneustes) fossilis,* in relation to body size. *Journal of Zoology,* **172,** 215–32.

Hughes, G. M. & Weibel, E. R. (1976). Morphometry of fish lungs. In *Respiration of Amphibious Vertebrates,* ed. G. M. Hughes, pp. 213–32. London & New York: Academic Press.

Hughes, G. M. & Wright, D. E. (1970). A comparative study of the ultrastructure of the water/blood pathway in the secondary lamellae of teleost and elasmobranch fishes – benthic forms. *Zeitschrift für Zellforschung und Mikroskopische Anatomie,* **104,** 478–93.

Koch, H. J. (1938). The absorption of chloride ions by the anal papillae of Dipteran larvae. *Journal of Experimental Biology,* **15,** 152–60.

Laurent, P. (1981). *Structure of vertebrate gills* (see this volume).

Lewis, S. V. & Potter, I. A. (1976). A scanning electron microscope study of the gills of the lamprey *Lampetra fluviatilis* (L). *Micron,* **7,** 205–11.

Marshall, N. B. (1960). Swimbladder structure of deepsea fishes in relation to their systematics and biology. *Discovery Report,* no. 31, 1–122.

Mill, P. J. (1972). *Respiration in the Invertebrates,* 212pp. London: Macmillan.

Miscalencu, D. (1973). Structure of the branchia in some oceanic teleosts. *Gegenbauers morphologisches Jahrbuch,* **119,** 449–53.

Morgan, M. (1974). Development of secondary lamellae of the gills of the trout, *Salmo gairdneri* (Richardson). *Cell and Tissue Research,* **151,** 509–23.

Muir, B. S. & Hughes, G. M. (1969). Gill dimensions for three species of tunny. *Journal of Experimental Biology,* **51,** 271–85.

Muir, B. S. & Kendall, J. T. (1968). Structural modifications in the gills of tunas and some other oceanic fishes. *Copeia,* **2,** 388–93.

Munshi, J. S. D. (1968). The accessory respiration organs of *Anabas testudineus* (Bloch) (Anabantidae, Pisces). *Proceedings of the Linnean Society, London,* **179,** 107–26.

Munshi, J. S. D. (1976). Gross and fine structure of the respiratory organs of air-breathing fishes. In *Respiration of Amphibious Vertebrates,* ed. G. M. Hughes, pp. 73–102. London & New York: Academic Press.

Nakao, T. (1976). The fine structure and innervation of gill lamellae in *Anodonta. Cell and Tissue Research,* **157,** 239–54.

Olson, K. R. (1979). The linear cable theory as a model of gill blood flow. *Journal of Theoretical Biology,* **81,** 377–88.

Pelseneer, P. (1935). *Essai d'Ethologic Zoologique,* 662 pp. *Academie Royale de Belgique.* Classe des Sciences, Publications Fondation. Brussels: Agathon Potter.

Piiper, J. & Baumgarten-Schumann, D. (1968). Effectiveness of O_2 and CO_2 exchange in the gills of the dogfish (*Scyliorhinus stellaris*). *Respiration Physiology,* **5,** 338–49.

Rahn, H. & Paganelli, C. V. (1968). Gas exchange in gas gills of diving insects. *Respiration Physiology,* **5,** 145–64.

Rankin, J. C. & Metz, J. (1971). A perfused teleostean gill preparation: vascular actions of neurohypophysial hormones and catecholamines. *Journal of Endocrinology,* **51,** 621–35.

Ruiter, L. de, Wolvekamp, H. P., Tooren, A. J. van & Vlasblom, A. (1952). Experiments on the efficiency of the 'physical gill' (*Hydrous piceus* L., *Naucoris cimicoides* L. & *Notonecta glauca* L.). *Acta Physiologica et Pharmacologica Neerlandica,* **2,** 180–213.

Scammell, C. A. (1971). Respiration and its nervous control in *Carcinus maenas.* Ph.D. Thesis. Bristol University.

Scammell, C. A. & Hughes, G. M. (1981). Comparative study of the functional anatomy of the gills and ventilatory currents in some British decapod crustaceans. *Biological Bulletin,* in press.
Scheid, P. & Piiper, J. (1976). Quantitative functional analysis of branchial gas transfer: theory and application to *Scyliorhinus stellaris* (Elasmobranchii). In *Respiration of Amphibious Vertebrates,* ed. G. M. Hughes, pp. 17–38. London and New York: Academic Press.
Singh, B. N. (1976). Balance between aquatic and aerial respiration. In *Respiration of Amphibious Vertebrates,* ed. G. M. Hughes, pp. 125–64. London and New York: Academic Press.
Smith, D. G. & Chamley-Campbell, J. (1981). Localisation of smooth-muscle myosin in branchial pillar cells of snapper (*Chrysophys auratus*) by immunofluorescence histochemistry. *Journal of Experimental Biology,* **215,** 121–4.
Steen, J. B. & Kruysse, A. (1964). The respiratory function of teleostean gills. *Comparative Biochemistry and Physiology,* **12,** 127–42.
Taylor, H. H. & Greenaway, P. (1979). The structure of the gills and lungs of the arid-zone crab, *Holthuisana* (*Austrothelphusa*) *transversa* (Brachyura:Sundathelphusidae) including observations on arterial vessels within the gills. *Journal of Zoology,* **189,** 359–84.
Thorpe, W. H. (1930). The biology, post-embryonic development, and economic importance of *Cryptochaetum iceryae* (Diptera, Agromyzidae) parasitic on *Icerya purchasi* (Coccidae, Monophlebini). *Proceedings of the Zoological Society of London,* **60,** 929–71.
Thorpe, W. H. (1932). Experiments upon respiration in the larvae of certain parasitic Hymenoptera. *Proceedings of the Royal Society of London, Series B,* **109,** 450–71.
Thorpe, W. H. (1950). Plastron respiration in aquatic insects. *Biological Reviews,* **25,** 344–91.
Tovell, P. W., Morgan, M. & Hughes, G. M. (1970). Ultrastructure of trout gills. *17th International Congres de Microscopie Electronique, Grenoble,* **3,** 601.
Vogel, W. (1978). Arterio-venous anastomoses in the afferent region of trout gill filaments. *Zoomorphologie,* **90,** 205–12.
Vogel, W., Vogel, V. & Kremers, H. (1973). New aspects of the intrafilamental vascular system in gills of a euryhaline teleost *Tilapia mossambica. Zeitschrift für Zellforschung und Mikroskopische Anatomie,* **144,** 573–83.
Weber, R. E. (1978). Respiration. In *Physiology of Annelids,* ed. P. J. Mill, pp. 369–92. London and New York: Academic Press.
Weibel, E. (1973). Morphological basis of alveolar–capillary gas exchange. *Physiological Reviews,* **53,** 419–95.
Wells, R. M. G. & Dales, R. P. (1975). Haemoglobin function in *Terebella lapidaria* L., an intertidal terebellid polychaete. *Journal of The Marine Biology Association, UK,* **55,** 211–20.
Wigglesworth, V. B. (1938). The regulation of osmotic pressure and chloride concentration in the haemolymph of mosquito larvae. *Journal of Experimental Biology,* **15,** 235–54.
Yonge, C. M. (1947). The pallial organs in the aspidobranch Gastropoda and their evolution throughout the Mollusca. *Philosophical Transactions of the Royal Society, B,* **232,** 443–518.
Youson, J. H. & Freeman, P. A. (1976). Morphology of the gills of larval and parasitic adult sea lamprey, *Petromyzon marinus* L. *Journal of Morphology,* **149,** 73–104.

PIERRE LAURENT

Structure of vertebrate gills

Introduction

Gills are the main respiratory organs of water-breathing vertebrates although skin (and in several groups the gas bladder) plays an additional but more or less facultative role. Aquatic breathers are mostly, but not exclusively, fish since larval and some adult forms of amphibians keep their gills functioning throughout life. In all these animals gills fulfil the same functions of gas transfer as do lungs for air breathers while in addition carrying out exchanges of mineral and organic water-soluble substances.

One of the most important consequences of having the gills surrounded by an aquatic medium is that, unless they are selectively permeable, or their permeabilities are strictly controlled, many soluble substances might pass right across the gill epithelia. An important characteristic of this medium is its unequal capacitances for the respiratory gases oxygen and carbon dioxide. The former is so poorly soluble in water that aquatic vertebrates have to handle considerable quantities of a dense and viscous medium passing over the gills in order to satisfy the oxygen requirement.

Recent developments in physiological studies on aquatic vertebrates, mostly fish, have shown clear trends in their functional organisation which explain, to a certain extent, how they are able to overcome some of the difficulties that water breathers have to withstand. These trends mainly concern a relative separation of gas and ionic exchange surfaces and the delicate organisation of the gill vasculature which allows efficient and controlled counter-current exchanges between blood and water. The aim of this present chapter is to give a short account of some characteristics of this functional organisation.

Gross anatomy

The present chapter deals with gills in the strictest sense, that is an internal, or in some cases an external, differentiation of the branchial septa that corresponds with the gill arches and lines two successive gill slits. It is well known that the structure and arrangement of gill arches depend upon the group considered, from the lower forms, e.g. the hagfish (Cyclostomata), up to the most highly evolved bony fish, e.g. the perch (Teleostei). For more details concerning this point, the reader should refer to basic zoology textbooks (Goodrich, 1930; Romer, 1970).

The present description of gill organisation mainly deals with the internal gills of fish*, but some complementary information on external gills will be given. External gills are not commonly present in adult fish but play an important role in Amphibian larval forms (tadpoles and neotenic larvae) and in water-breathing adults (e.g. *Necturus maculatus*).

The gill apparatus consists of three components: the gill arch supports the primary lamellae (sometimes referred to as filaments) and these in turn bear the secondary lamellae which are the sites of gas exchanges. Gill arches and primary lamellae comprise different pieces of skeleton; the so-called gill bars in the for-

Fig. 1. Scanning electron micrograph of a primary lamella of *Solea solea*. On the afferent side (af) small pits reveal chloride cells. Note that they are absent on the efferent side (ef). The arrow marks the direction of water flow between the secondary lamellae (sl).

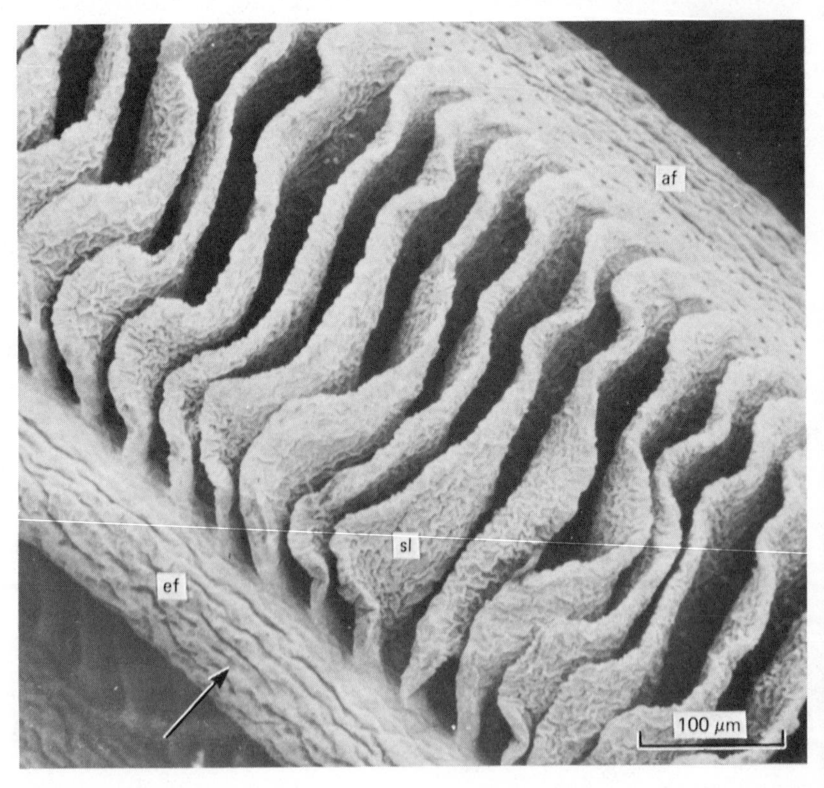

*Throughout this text, fish is used in its widest sense, i.e., including all classes from cyclostomes up to teleosts.

mer and gill rays in the latter. They are actively controlled in the course of the breathing events by a system of striated and smooth muscles. By this means the secondary lamellae are correctly positioned in the respiratory water flow. This basic disposition is seen in all groups of fish with the exception of some dipnoans, of which *Protopterus* and *Lepidosiren* will be considered separately.

Two rows of primary lamellae are generally inserted on each gill arch and secondary lamellae are regularly spaced on primary lamellae (Fig. 1). This whole forms the so-called holobranch. In some groups (Chondrostei and Holostei) certain arches (hyoidean and mandibular) only bear a single row of primary lamellae; they are called hemibranchs. The mode of insertion of primary lamellae on gill arches varies somewhat with the group considered. For instance in the Chondrichthyes, septa prolong the inner side of gill arches. The septa delimit branchial pouches which are lined on both sides by primary lamellae fixed at their internal edge. This pattern regresses in the higher groups, so that gill primary lamellae become progressively freer. In Teleostei they are only fused at their base. Whatever the mode of arrangement, the respiratory flow of water crosses the primary lamellae in a counter-current direction with the blood.

Gill arches and primary lamellae distribute the blood flow into the secondary lamellae. The gill vasculature is made up of afferent arteries of three successive orders; branchial arteries (in the gill arch itself), primary arteries (in the axis of primary lamellae and sometimes referred to as filamental arteries) and secondary arteries (or arterioles) which supply the capillary network of the secondary (or respiratory) lamellae. After its passage into these capillaries, the blood is collected by an identical but efferent system of arteries and finally sent out into the dorsal aorta towards the systemic bed. Arterial walls are generally thicker on the afferent side, probably because of the higher blood pressure which prevails on this side.

In each gill arch there is a large vein, the branchial vein, which is connected with the jugular vein. Branchial veins collect the blood from the small veins of the primary lamellae. Thus two interconnected blood pathways, arterio–arterial and arterio–venous, form the gill vasculature.

The arterio–arterial blood vasculature (*Fig. 2*)

The afferent branchial artery gives off, at intervals, afferent primary arteries (aa) and from these short afferent secondary arteries (or arterioles) supply the secondary lamellar vascular network. On the efferent side of this vasculature, a sphincter (sph_1) controls the blood flow just before its entrance to the efferent branchial artery. This sphincter consists of a reinforcement of the muscle layer which is several times thicker than the efferent artery. Efferent primary sphincters are innervated by numerous nerve endings filled with small cholinergic vesicles. This area shows a strong positive histochemical reaction for cholinesterase.

Fig. 2. Schemata of the gill vasculature of (*a*) *Salmo gairdneri* (Teleostei) and (*b*) *Acipenser baeri* (Holostei). On the left part of each schema: gill arch in cross section; on the right, the upper hemibranch shows arterio–venous system, the lower hemibranch shows arterio–arterial vasculature of primary and secondary lamellae in sagittal view. aa, afferent artery; aba, afferent branchial artery; ava_{af}, arterio–venous anastomosis on the afferent side; ava_{ef}, arterio–venous anastomosis on the efferent side; bv, branchial vein; c, cartilage; ci, cisterna; cc, corpus cavernosum; cvs, central venous sinus; ea, efferent artery; eba, efferent branchial artery; es, extracellular spaces; m, muscle; n, nerve; na, nutrient artery; pl, primary lamella; sh, arterio–arterial shunt; sl, secondary lamella; sph_1, efferent artery sphincter; sph_2, pre- or post-lamellar sphincter. (From Dunel & Laurent, 1980.)

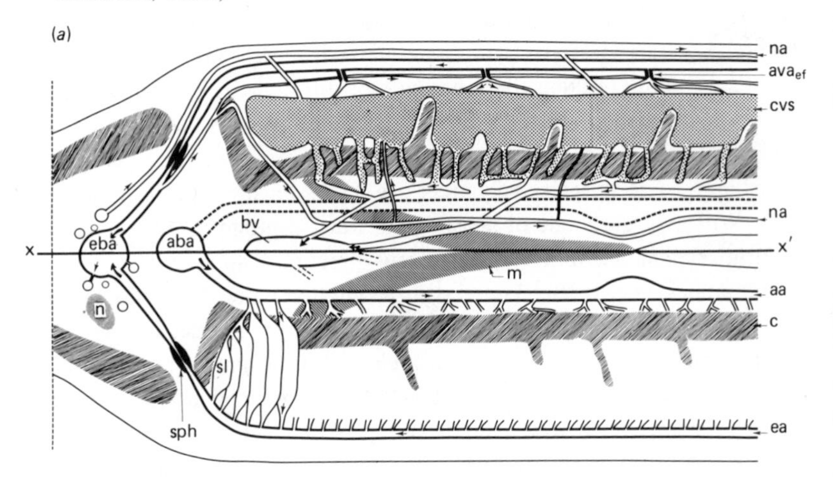

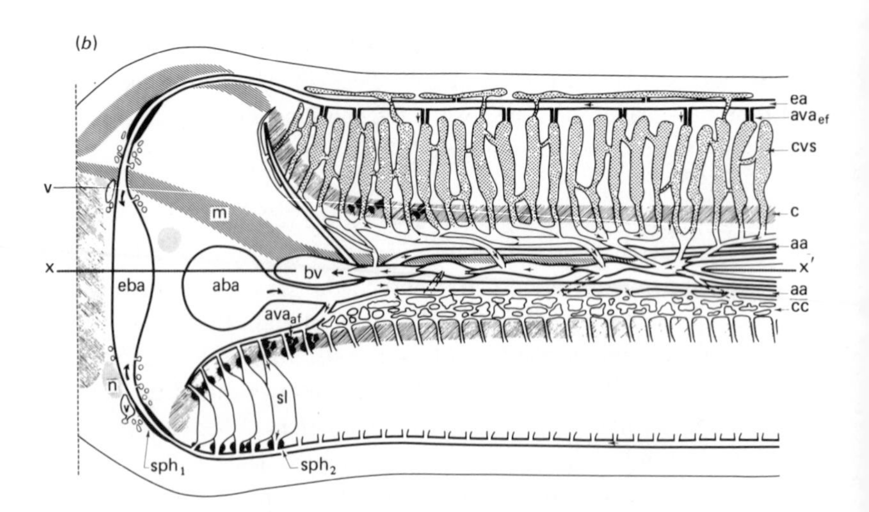

In certain groups of non-teleost fish, such as the sturgeon (Chondrostei) and the dogfish and sharks (Chondrichthyes), short secondary arteries supply the secondary lamellae through a mass of cavernous tissue (Fig. 3, right). This tissue forms a loose network of intermingled channels of various sizes. In the skate *Raja*, the cavernous tissue forms a large sponge-like system which, when filled with blood, stiffens the primary lamellae and presumably favours an erected position of the secondary lamellae.

Another important component of the arterio–arterial vasculature is the presence of sphincter-like structures (sph_2) located on both afferent and efferent secondary arteries. These sphincters control both secondary lamellar inflow and outflow. Ultrastructural studies reveal the presence of a well-delineated smooth muscle ring consisting of several circular layers. The presence of a specific innervation is still questioned although numerous nerve fibres and ganglionic cells are visible in their vicinity. Secondary sphincters, as they are called, are particularly obvious in non-teleost fish, such as the sturgeon, and several Chondrichthyes, such as the dogfish (Fig. 3, left). In teleosts secondary sphincters are not so obvious but the secondary lamellae possess potent contractibility as shown in cast preparations, suggesting a functional if not morphological specialisation of the arterial walls. Such sphincters have also been found in Holostei (the garpike and the bowfin).

In the lungfish (Dipnoi, *Protopterus* and *Lepidosiren*), the arterio–arterial system is completely different from that in other classes of fish; the gills do not form

Fig. 3. Cast of the primary lamella vasculature of *Scyliorhinus canicula*. On the right, afferent side with cavernous tissue (ct) interposed between afferent primary artery (aa1) and secondary lamellae (sl). On the left, efferent side showing sphincters (sph_2) on secondary arteries. Arterio–venous anastomoses (ava) are visible between efferent primary artery (ea1) and central venous sinus (not shown).

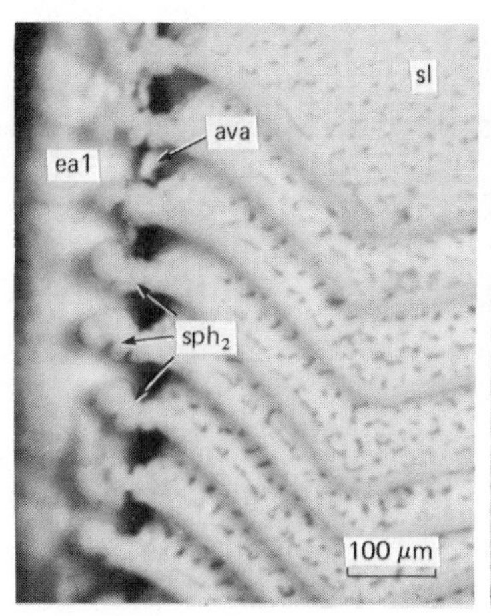

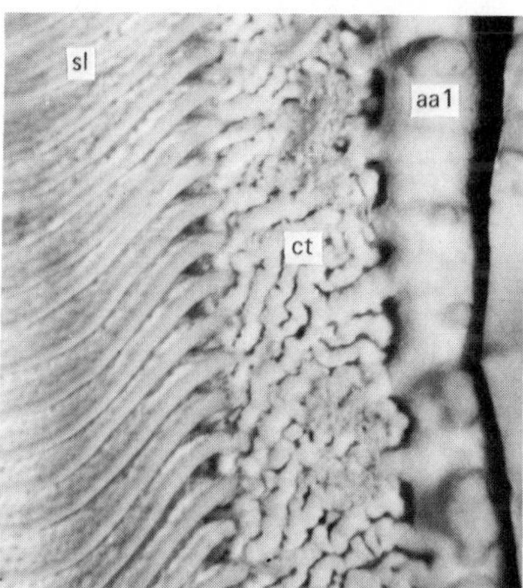

regular arrays of primary and secondary lamellae, but resemble the external arborescent gills of a tadpole rather than the gills of Teleostei and Chondrichthyes.

As shown on the schema of Fig. 4, a series of afferent arteries (aa) emanating from the branchial artery gives off secondary, tertiary and even quaternary arteries which supply the lamellae of corresponding order. A particular and functionally important feature is the presence of wide, short and direct channels (sh) connecting the lamellar afferent and efferent arteries near to their junction with the corresponding branchial arteries. Thus, efficient gill shunting pathways, which are never observed in other classes of fish, make the gill vasculature of the Dipnoi particularly interesting. It is worth mentioning that a similar arrangement characterises the vasculature of the external gills in *Amblystoma* (axolotl) larvae.

From a functional viewpoint, secondary lamellae are, of course, the predominant part of the arterio–arterial system (Fig. 5). Briefly speaking, a secondary lamella consists of a flat capillary lined on both of its sides by an epithelium which is anchored on pillar cells whose flanges form the endothelium. Pillar cells are another fundamental feature of fish gill capillaries. They are endothelial in function since they line this part of the blood compartment. It has been sug-

Fig. 4. Schema of the gill vasculature of *Protopterus aethiopicus* (Dipnoi). See abbreviations on caption of Fig. 2.

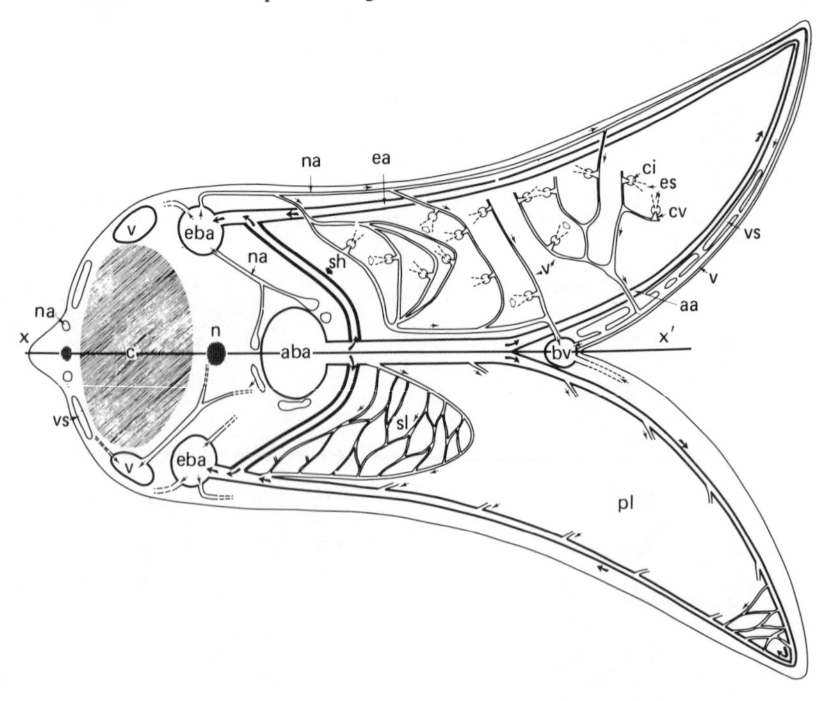

gested, according to their situation, that they might contract or dilate and by this means give rise to large changes in the lamellar blood flow. Structurally, they display intracytoplasmic bundles of actomyosin filaments which when contracted might reduce the lumen width. Moreover, they are reinforced by columns of collagen which might work as slide-bars.

Three kinds of secondary lamellae can be found among the different classes of fish.

The teleost type is characterised by a polygonal distribution of pillar cells (Fig. 6). This probably represents the most efficient system for exchanging gas, since blood must flow through a system of chicanes. This type is also found in Chondrostei and Chondrichthyes, so that it is, by far, the most common. In *Acipenser* for instance, two marginal blood vessels, an inner and an outer, are conspicuously differentiated from the rest of the lamellar network. They taper off in opposite directions, suggesting that the inner vessel distributes the blood inflow through the lamellar capillaries and the outer vessel gathers the blood outflow.

The holostean type is observed in *Lepisosteus* and *Amia*. Pillar cells are arranged in parallel arrays and are fused side by side. This arrangement is a compromise between maximum respiratory efficiency and lowest dynamic resistance. Such an arrangement could explain how large variations in gill resistance can be brought about in those dual breathers during air-breathing periods, when gills are useless.

The dipnoan type is completely devoid of pillar cells. The vasculature is made up of a loose network of large capillaries which accommodate the large size of the erythrocytes and extend between the afferent and efferent arteries.

Fig. 5. Electron micrograph of a cross-section of a secondary lamella of *Salmo gairdneri*. pvc, pavement cell; ndc, non-differentiated cell; pc, pillar cell (with its flanges); col, collagen column; cap, capillary; bl, basal lamina; lym, lymphocyte; sp, intercellular (lymphoid) space; em, external milieu.

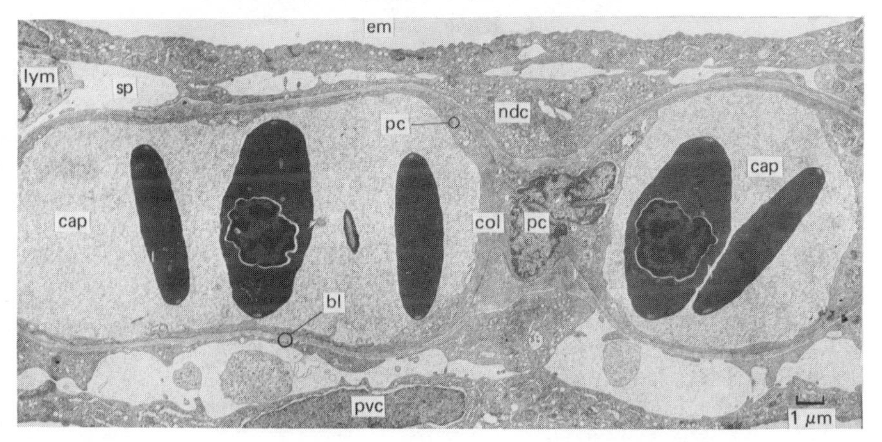

The arterio–venous blood vasculature (*Fig. 2*)

The venous circulation has recently been the focus of some interest. Arterio–venous components form two systems. The first one is functionally nutritive and consists of small arteries which take their origins from efferent branchial and primary arteries. After variable courses in the gill parenchyma, nutrient vessels collect into branchial veins or their tributaries. A second system more recently studied is probably not nutritive and mainly consists of: (i) the so-called central venous sinuses of the primary lamellae; (ii) a number of arterio–venous anastomoses supplying the central venous sinuses from the arterio–arterial blood pathways; (iii) the collecting venous pathways.

Fig. 6. Cast of the primary lamella vasculature of *Acipenser baeri* cross-sectioned to show secondary lamellae. The capillary system of secondary lamellae (sl) is filled with cast material and pillar cells are represented as small black areas. Note the large size of the central venous sinus (cvs) and its extensions, especially around the primary efferent artery (ea1) and cartilage (c).

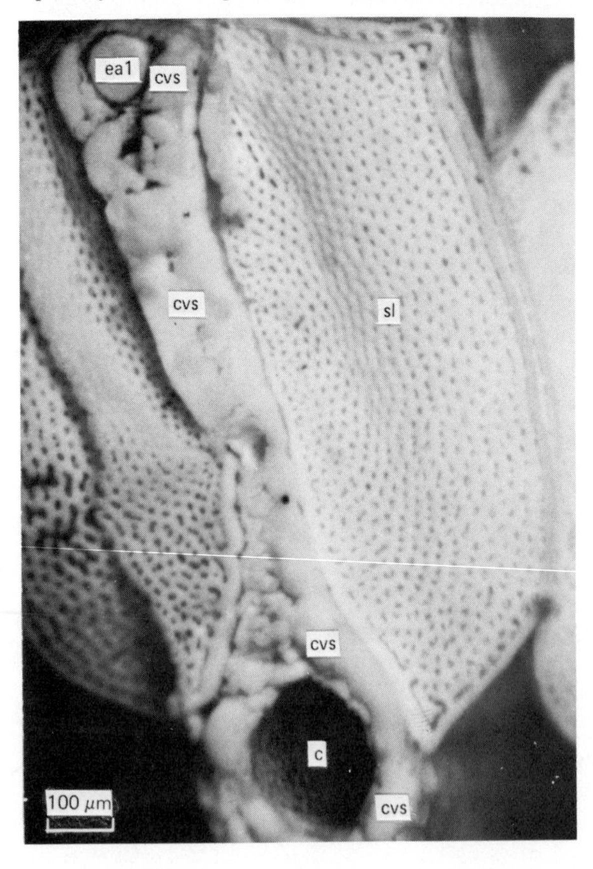

The location of the anastomoses varies according to the species considered. In Teleostei, the central venous sinus communicates with the efferent primary arteries, except in a few cases such as the eel and certain catfish, which display anastomoses from both afferent and efferent arteries. This difference is perhaps of some importance since, in the second case, some non-oxygenated blood passes into the venous compartment.

The sturgeon (Chondrostei) displays anastomoses on both sides as does *Lepisosteus* (Holostei). Thus it appears that in the most primitive forms of teleost, as well as in the lowest groups of fish, anastomoses are not restricted to the efferent side. In *Amia* (Holostei), this situation is even more complicated because of an arterial supply to the central sinus which originates directly from branchial arteries in addition to that from the primary arteries. In some Chondrichthyes, for example *Raja*, anastomoses are also located on both sides.

The basic structural characteristics of these anastomoses are as follows: (i) the anastomoses are rather short channels, generally less than 100 μm; (ii) the anastomosis wall is thicker than that of an artery of the same size, owing to the presence of from four to six layers of smooth muscle; (iii) their lumens are generally occluded by endothelial cells, which leads one to think that anastomoses are generally fixed for histology in a state of closure.

Smooth-muscle fibres and endothelial cells are often in close apposition through interruption of the basement membrane (nexus), a disposition which could explain a particular reactivity to luminal stimuli. As far as one knows, anastomoses are not provided with nerve endings and are presumably under a blood-borne humoral control as suggested by recent experiments. However, large nerve endings (5 μm in diameter) filled with dense-cored vesicles are found within 0.5 μm of smooth-muscle fibres in the dogfish, suggesting a local control by chromaffin cells.

The central venous sinus is a hollow compartment of varying complexity and occupies the full length of the primary lamellae (Figs. 6 and 7). It is wider in freshwater fish than in seawater species as far as one can tell by estimating this dimension from specimens fixed for histology. In *Anguilla*, the central sinus looks like a simple sac with extensions overlapping afferent and efferent primary arteries. It is far more complex in *Perca* and in many other species where a branched system of cavities extends finger-like appendages, often very superficially, and surrounds afferent and efferent arteries beneath the epithelium. The greatest complexity seems to be achieved in *Amia*. Many of these organisational patterns suggest that the central venous processes correspond with the interspace between the secondary lamellae.

In Dipnoi, the arterio–venous vasculature differs from that of the other classes by lacking a central venous sinus. A typical venous system comes from the

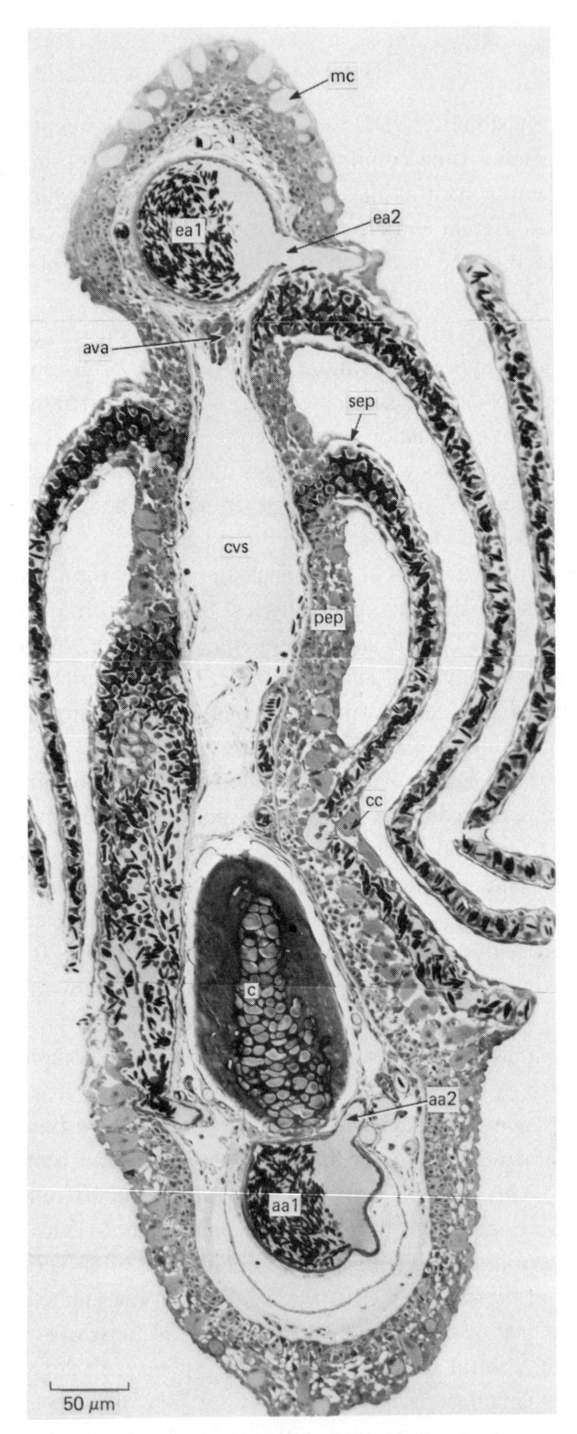

Fig. 7. Semi-thin cross-section of a primary lamella of *Anguilla anguilla*. Note both the disposition of the primary (pep) and secondary (sep) epithelia and the size of the central venous sinus (cvs). Arterio–venous anastomoses (ava) supply

nutrient capillaries and finally drains into a branchial vein that runs in a parallel direction to the afferent branchial artery.

From an ultrastructural viewpoint, the central venous sinuses have a vein-like structure with a low ratio of wall thickness to lumen width. Their walls are lined by a thin endothelium which is poor in muscle fibres and elastin and therefore presumably compliant.

The central venous sinuses communicate with the venous drainage by different means; either by veins running parallel to the afferent primary arteries, as in the trout or the perch, or by a direct connection with the branchial vein, as in the eel.

Gill epithelia (*Fig. 7*)

Two types of epithelia are generally encountered when examining the gill surface. Distinction between them is based on differences bearing both on their location and on their well-defined structure and pattern of vascularisation. The primary epithelium surrounds the whole gill apparatus except the secondary lamellae, which are covered by the so-called secondary epithelium.

The primary epithelium

This epithelium is multilayered and surrounds all the primary lamellae, threading its way between the regularly-spaced secondary lamellae. Hence the primary epithelium is also called the interlamellar epithelium, a term which is too restrictive because of the broader distribution of this epithelium. The vascular compartment associated with the primary epithelium consists of the central venous sinus and its extensions in the primary lamellae. It is noteworthy that in the branchial arch itself, as in the primary lamellae, blood sinuses are located underneath the arch epithelium, which has characteristics similar to those of the primary epithelium.

The main feature of the primary epithelium is the presence of chloride cells. These cells are mainly located in the interlamellar regions where they predominate, either scattered or in clusters (Fig. 8). They often spread up to the base of secondary lamellae. In marine species, or in euryhaline seawater-adapted species such as the eel, where chloride cells are very numerous, they are also located side by side in the non-lamellar area, namely along the afferent edge of the primary lamellae. Extrabranchial locations have also been reported, for instance among the cells lining the opercular cavity. Chloride cells are in close association

the central venous sinus. Chloride cells (cc) are visible within the primary epithelium, preferentially on the afferent side of the primary lamella. Note the mucous cells (mc) on the efferent side. c, cartilage; aa2, secondary afferent artery (arteriole); ea2, secondary efferent artery (arteriole). (From Laurent & Dunel, 1980.)

with pavement cells, with which they form the outermost layer of the epithelium. Pavement cells are either columnar, displaying a large number of cytoplasmic vesicles and deeply infolded parietal membranes, or squamous, displaying an extensive Golgi system. The basal layer of this stratified epithelium consists of typical non-differentiated cells.

The main ultrastructural characteristics of chloride cells include their richness in mitochondria which are associated with a densely branched tubular system that opens on the basolateral plasma membrane. The apex of each cell has microvilli and is firmly bound to the neighbouring pavement cells by a long and tight junctional complex. A terminal web is anchored by parietal desmosomes. Vesicles of various sizes are concentrated in the apical region. Chloride cells are present in both euryhaline and stenohaline species of fresh or saltwater teleosts and exhibit specific, significant changes in relation with the medium. In seawater or seawater-adapted fish, the apical membrane of chloride cells is more or less deeply buried at the bottom of a crypt consisting of an opening through the pavement cell layer. In freshwater fish, the apical surfaces of chloride cells are more broadly exposed to the external medium. But the most fascinating differ-

Fig. 8. Scanning electron micrograph of the gill epithelia of *Anguilla anguilla*. On the primary epithelium (pep) note the apical membranes of the chloride cells (cc). Part of a secondary lamella visible at the top displays the more ridged surface of its epithelium (sep).

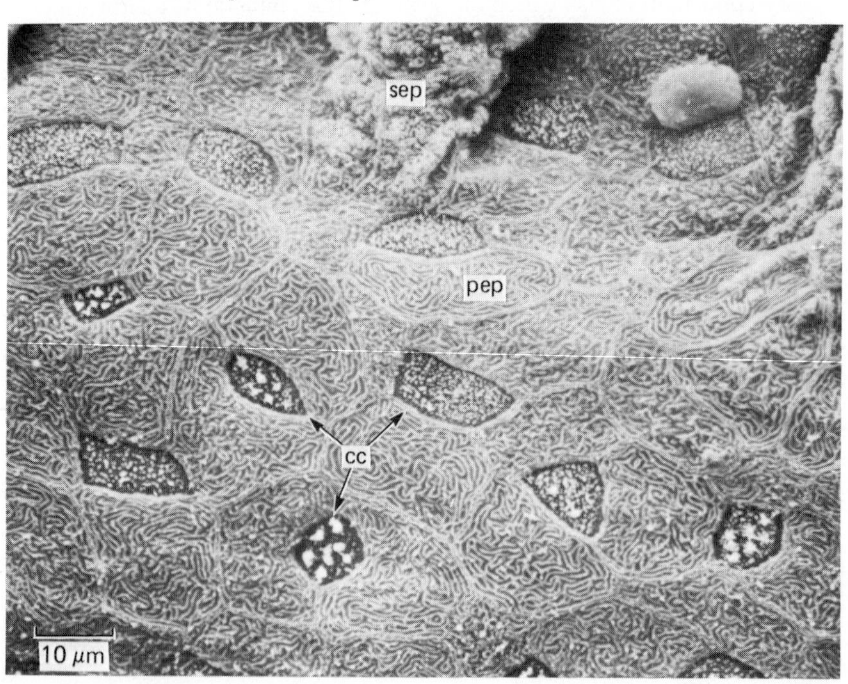

ence is the presence of an accessory cell beside each chloride cell in saltwater fish. Accessory cells have never been reported in freshwater fish. Cytoplasmic processes of the accessory cell interdigitate with the apical chloride cell cytoplasm (Fig. 9). Consequently, the apical surface of this complex is a mosaic arrangement of the two cell membranes. The two membranes are fastened together by a specific very short junctional apparatus which differs from the tight junctional apparatus characteristic of the freshwater-fish primary epithelium. This intercellular junction is presumed to be leaky on the basis of freeze-etching data. This means that the primary epithelium, which is tight in freshwater fish, becomes leaky in seawater fish, a structural difference which could have important functional implications for osmoregulation.

The dynamics of development of the accessory cells when a euryhaline fish is transferred from freshwater to seawater has not yet been elucidated. First of all, there is little doubt that accessory cells are not juvenile chloride cells, as suggested by some authors, but constitute a definite form with a definite function. The second point is that accessory cells develop rapidly after experimental trans-

Fig. 9. Scanning electron micrograph of a chloride cell (cc) of *Solea solea*. Its apical membrane is buried between pavement cells (pvc). Note the apical microvilli (mv) and the cytoplasmic processes of an accessory cell (ac).

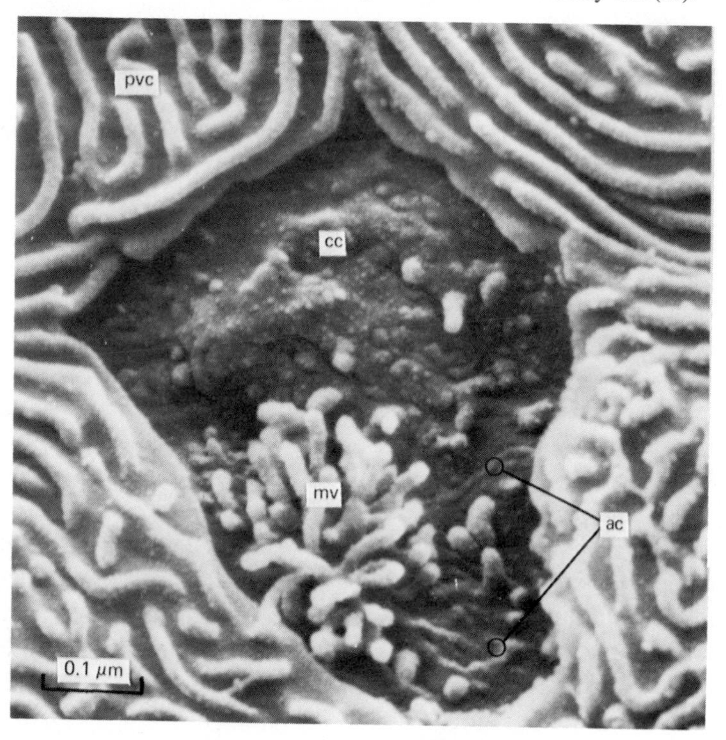

fer from freshwater to seawater, for instance the freshwater eel displays accessory cells after adaptation to 50% seawater for one week and three days' adaptation is sufficient at 100% seawater. Nothing is yet known about the earliest stages of development. Whether or not the unique function of accessory cells is to allow epithelial leaky junctions is still a matter of speculation.

Chloride cells are also present in non-teleost fish gills although they often display a somewhat different structure. In Chondrostei (the sturgeon) and Holostei (the garpike and the bowfin), chloride cells have a morphology which is the same as in teleosts.

In Chondrichthyes, two types of mitochondria-rich cells have been found, and since both of them have relations with the external medium via an apical membrane, they have been considered as chloride-cell-like (Fig. 10). In the first type, the apical membrane is deeply buried in a cul-de-sac and communicates with the external medium by a narrow opening. In the second type, the apical membrane protrudes outward. Both types lack the tubular system so conspicuous in teleost chloride cells, but intracellular tubules are replaced by numerous basolateral membrane infoldings which presumably fulfil the same function. It is interesting to note that numerous mitochondria are associated with the infoldings, a feature which is a constant characteristic of several types of cells with a transport function.

In cyclostomes, the chloride cells display a number of characteristics found in teleosts. In the sea lamprey *Petromyzon marinus*, there is a close similarity between the chloride cell of the freshwater-adapted adult lamprey and those of marine teleosts, indicating a preadaptation to a future exposure to a marine environment. In contrast, the freshwater larval form (ammocoete) possesses only a few chloride cells which are structurally similar to those of freshwater teleosts.

In Dipnoi, the poorly developed arborescent gills have only a single type of epithelium which is not well adapted for gaseous exchanges, but may be active in hydromineral exchanges. In *Protopterus,* as well as in *Lepidosiren,* the gills display a thick multilayered epithelium made up of cells containing numerous small round mitochondria. The outermost cells have microvilli at their apical surface and are linked together by long, convoluted junctional apparatuses which give rise to large intercellular channels filled with interdigitating processes of the basolateral membranes. Thus these cells have many of the structural characteristics of ion-transporting cells.

Mucous cells are numerous on the gills, and are predominantly located within the primary epithelium, principally on its efferent side. They also extend onto the gill arches themselves. There is no clear evidence suggesting the mucous cells are more or less abundant in any particular conditions that the fish has to cope with, though they discharge in response to violent changes in the medium (a decrease in pH, for instance).

Fig. 10. Electron micrograph of the primary epithelium of *Scyliorhinus canicula*. A large chloride-like cell (cc) with its numerous mitochondria and basolateral infoldings rests on the wall of the central venous sinus (cvs). Pavement cells (pv) line the chloride-cell pit and form the boundary with the external medium (em).

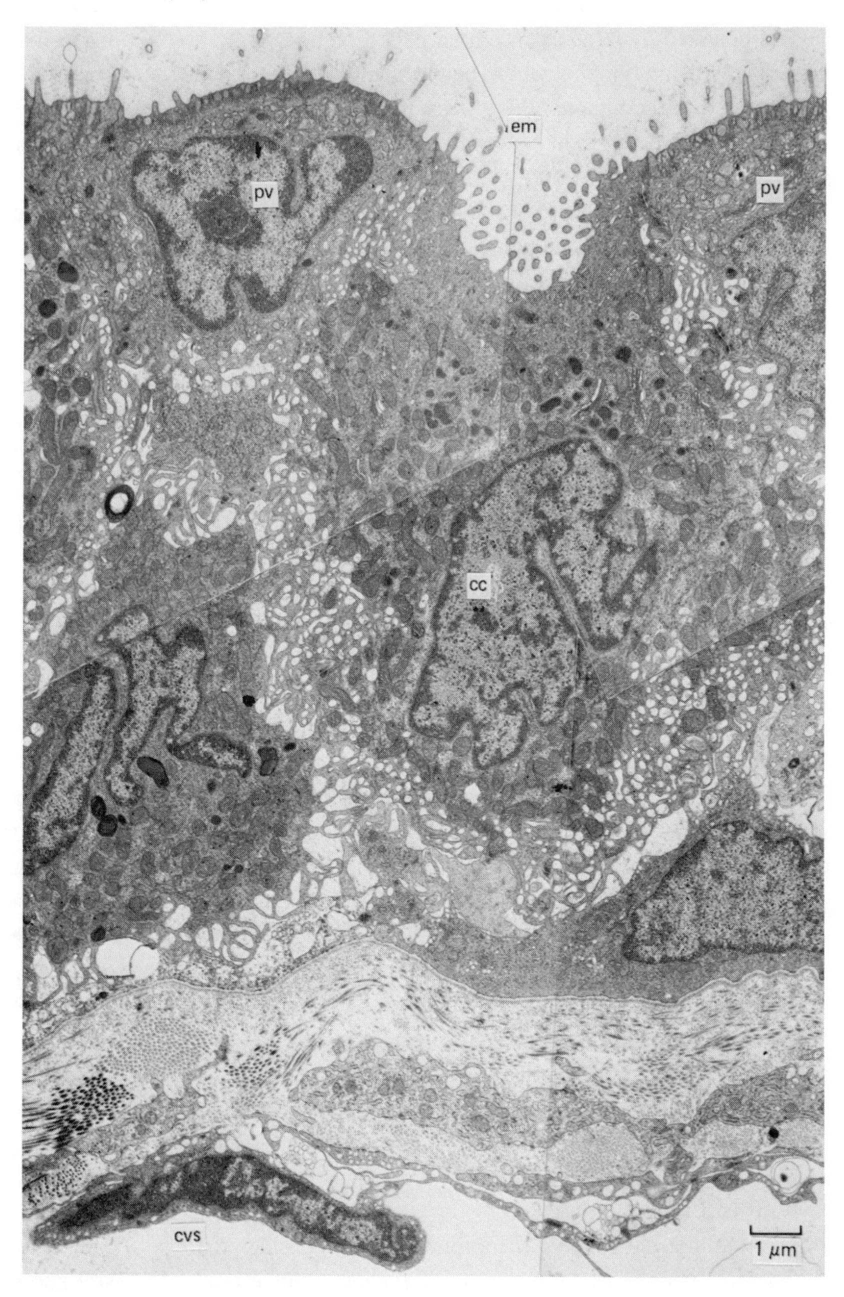

The secondary epithelium (Fig. 5)

The epithelium of secondary lamellae has a structure which is similar throughout the different groups of fish except Dipnoi, as already mentioned. It rests on the basal lamina of pillar cells. This epithelium is made up of two cell layers separated by large extracellular gaps which, at some places, may be wide enough to contain lymphocytes. They form a lymphatic-like network which communicates with the intercellular spaces of the primary epithelium and the venous sinuses. This feature might have some implications in consideration of transbranchial fluid movements possibly draining into the low pressure vasculature.

The external layer of the secondary epithelium, also called the mucosal layer, is characterised by its pavement cells. These cells are highly differentiated since they possess a well-developed Golgi apparatus, an abundant rough endoplasmic reticulum and vesicles of various sizes. These characteristics point to an important cellular activity, the functional significance of which is still unknown. Among its specific properties, the gill epithelium must be as permeable as possible to gases and relatively less permeable to ions and water, especially for saltwater fish. Because of the large surface area occupied by the secondary epithelium, these cells probably contribute the major part of the passive permeability processes. So far there is no evidence regarding changes in the secondary epithelium ultrastructure associated with different external conditions. Nevertheless, and in spite of the absence of obvious differences, rearrangement of membrane phospholipid molecules might cause the epithelium permeability to change. On the other hand, gill permeability might be affected by the composition of the cellular coat surrounding the external epithelial boundary surface.

The innermost layer consists of typical non-differentiated cells with a high nuclear to cytoplasmic volume ratio. However, numerous ribosomes and a well-developed rough endoplasmic reticulum suggest that some of them start to differentiate. Some observations suggest that these cells develop and give rise to chloride cells in some circumstances, though normally chloride cells originate, as mentioned above, in the primary epithelium. These circumstances appear to be related to abnormal conditions inducing an acute loss of ions, like skin wounds, fungal diseases or a stay in deionised, calcium-free water. Then an exaggerated development of chloride cells takes place in the secondary lamellae, presumably providing a compensatory mechanism for increasing the net ion uptake. It has been noticed that, in these conditions, these cells have an abnormally large contact area with the external medium, which quickly decreases when the animals are returned to normal conditions.

Gill morphometry

As for the lung, gill morphometry is considered to be a tool which allows the physiologist to evaluate gill exchange potentiality. The well-known

Fick's law of diffusion defines the diffusion rate of a gas, for instance oxygen, as being proportional to the exchange surface area and inversely proportional to the thickness of the diffusion barrier. That is why some efforts have been made during the last decade to relate these morphometric values to size of gill, oxygen consumption, species, behaviour or external conditions. As a general rule, it can be said that the gill area, that is, the product of the number of secondary lamellae and their mean surface area, follows an allometric law when correlated with the body weight or oxygen consumption in the same species. In the same way, it has been observed that tuna, one of the most active fish, displays a greater area per unit of body weight than any other species. This value is close to 2 m^2 for an animal weighing 1 kg. A much less active fish, for instance the small mouth bass of the same weight, has a total secondary lamellar area which does not exceed 0.13 m^2. The observed difference is due to an increased number of secondary lamellae (tuna) or (and) an increased mean surface area of the secondary lamellae. It has also been shown that chronic hypoxia triggers a progressive increase of secondary lamellar area. For instance, in catfish (*Ictalurus melas*) weighing 100 g and kept in weakly-oxygenated water for six weeks, the mean area is in the range of 0.23–0.27 mm^2 per lamella, values which are about twice those found under normal conditions (range 0.11–0.15 mm^2). It is interesting to note that this increase relates to the free dimension of the secondary lamella, in other words the secondary lamellae increase their length. In the same conditions of hypoxia and in the same animals, the secondary epithelium harmonic mean thickness comes down to 3 μm in contrast with the normoxic conditions where thickness ranges between 6 and 7 μm. Those few examples emphasise the extent to which gill epithelia are adaptable structures and how much these adaptations are functionally significant since, after chronic exposure to hypoxia, the morphometrically calculated overall gill gas transfer rate per unit of partial pressure difference is four times greater than in normoxia.

Amphibian gills

In amphibians, a distinction is currently made between external and internal gills. There are some differences between them when their embryological development is taken into account. Nevertheless, they have the same function during aquatic life. Primitive Tetrapoda inherited gills from their fish ancestors. By this time, four branchial arches bearing gills covered by an operculum made up their respiratory apparatus. In the modern tadpole larvae of Anura, internal gills form two rows of branched lamellae supported by the bars separating the four gill slits. Later on, when pulmonary respiration takes place, the gill slits close, except in some Urodela which readapted to an aquatic life and keep some slits open.

External gills develop earlier and are of two kinds. The first, in larval forms of elasmobranchs, form long filaments floating in the albuminous fluid within the egg case. Some larvae of Chondrostei and Teleostei also have such gill filaments. The second, which are true external gills, occur in the larval stages of Dipnoi, Polypterini and Amphibia. Urodela readapted to an aquatic life keep their external gills functioning throughout their life. Another interesting example is the neotenic larva of *Amblystoma* (Axolotl) which also has external gills serving as a perennial respiratory organ. Amphibian external gills differ from those of fish in many ways; macroscopically they form arborescent organs and are never arranged in well-ordered lamellae of first and second order.

In the axolotl, the gill vascular circuit is built as a derivation of the aortic arch proper. This secondary circuit is composed of afferent branchial arteries giving off a series of tributaries or afferent lamellar arteries. Each lamellar artery forms capillary networks. A symmetrical efferent system collects the blood which passes into the efferent branchial arch. It is noteworthy that the blood can also pass through from the afferent to the efferent side of the gill arch by a series of anastomotic arterioles (Fig. 11). When these arterioles contract, the blood is forced into the gill vasculature. Interestingly, adrenaline causes the anastomoses to contract and the lamellar capillaries to dilate. The same antagonistic action is revealed by oxygenated blood. It should be mentioned that Dipnoi display the same pattern of gill organisation by allowing the blood either to perfuse the gill lamellae or pass straight towards the pulmonary vascular bed.

In the axolotl, the gill epithelium is of a single multilayered type. Underneath this epithelium is a double capillary network irrigating both faces of the lamella.

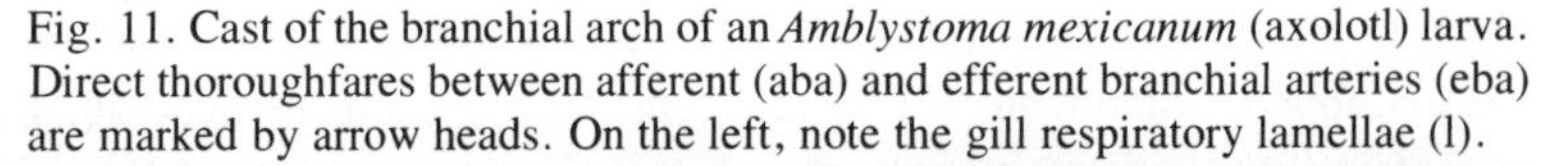

Fig. 11. Cast of the branchial arch of an *Amblystoma mexicanum* (axolotl) larva. Direct thoroughfares between afferent (aba) and efferent branchial arteries (eba) are marked by arrow heads. On the left, note the gill respiratory lamellae (l).

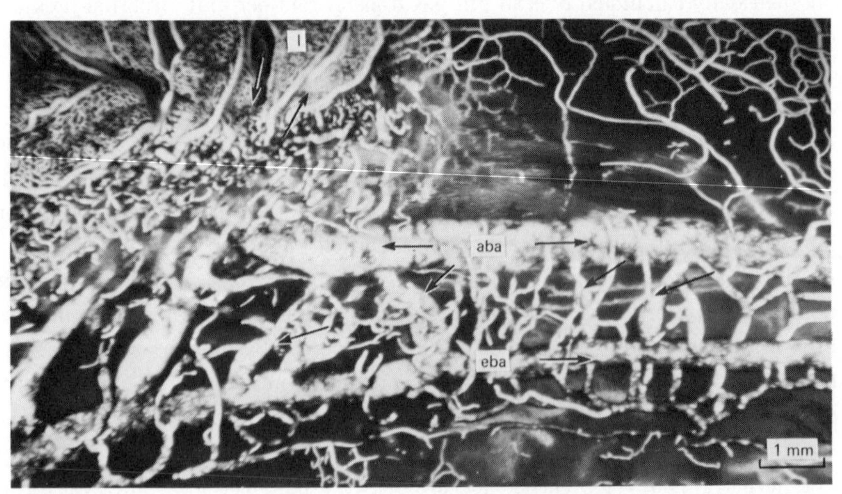

In the inner part, large thin-walled vessels presumably represent a venous compartment. With the electron microscope, the outermost layer of the multilayered epithelium can be seen to consist of several types of cells; one of them cuboidal in form and rich in mitochondria, displays prominent Golgi complexes giving rise to numerous vesicles of different sizes. It is also characterised by an endocellular skeleton of thin filaments. All these characteristics indicate a transporting epithelium, so that the cell type has been considered, on the basis of its morphological characteristics, as the equivalent of teleost chloride cells. Similar conclusions can be drawn from the structure of internal gills in the *Discoglossus* larva where abundant populations of transporting cells are also found.

References

The list of publications given below is far from being an extensive bibliography on gill morphology which counts several hundred references. The reader should find numerous quotations in the listed publications or might personally ask the author of the present chapter.

Dunel, S. & Laurent, P. (1980). Functional organisation of the gill vasculature in different classes of fish. In *Epithelial Transport in the Lower Vertebrates*, ed. B. Lahlou, pp. 37–58. Cambridge University Press.

Dunel-Erb, S. & Laurent, P. (1980). Ultrastructure of marine Teleost gill epithelia: SEM and TEM study of the chloride cell apical membrane. *Journal of Morphology*, **165,** 175–86.

Figge, F. H. J. (1936). The differential reaction of the blood vessels of a branchial arch of *Amblystoma tigrinum* (Colorado Axolotl). I. The reaction to adrenaline, oxygen and carbon dioxide. *Physiological Zoology*, **9,** 79–101.

Goodrich, E. S. (1930). *Studies on the Structure and Development of Vertebrates*. New York: Dover Publications Inc, 1958.

Hackford, A. W., Gillies, C. G., Eastwood, C. & Goldblatt, P. J. (1977). Thyroxine-induced gill resorption in the Axolotl (*Amblystoma mexicanum*). *Journal of Morphology*, **153,** 479–504.

Houdry, J. (1974). Etude des branchies 'internes', puis de leur régression au moment de la métamorphose chez la larve de *Discoglossus pictus* (Otth), Amphibien anoure. *Journal de Microscopie*, **20,** 165–82.

Hughes, G. M. (1972). Morphometrics of fish gills. *Respiration Physiology*, **14,** 1–25.

Laurent, P. (1980). Circulatory adaptation to diving in amphibious fish. *Proceedings of the International Union of Physiological Sciences*, **14,** 175 (abstract).

Laurent, P., Delaney, R. G. & Fishman, A. P. (1978). The vasculature of the gills in the aquatic and aestivating Lungfish (*Protopterus aethiopicus*). *Journal of Morphology*, **156,** 173–208.

Laurent, P. & Dunel, S. (1976). Functional organisation of the teleost gill. I. Blood Pathways. *Acta Zoologica* (Stockholm), **57,** 189–209.

Laurent, P. & Dunel, S. (1980). Morphology of gill epithelia in fish. *American Journal of Physiology*, **238,** R147–59.

Romer, A. S. (1970). *The Vertebrate body*. Philadelphia: Saunders Co.

J.PIIPER & P.SCHEID

Physical principles of respiratory gas exchange in fish gills

Introduction

The aim of this chapter is to attempt an analysis of gas exchange in fish gills by the use of relatively simple models. The study is an extension of previous investigations on models for gas exchange in various respiratory organs of vertebrates (Piiper & Scheid, 1972, 1975) and on gas transfer in fish gills (Scheid & Piiper, 1971, 1976). Measurements performed on the elasmobranch *Scyliorhinus stellaris* (Piiper & Schumann, 1967; Baumgarten-Schumann & Piiper, 1968; Piiper & Baumgarten-Schumann, 1968*a, b;* Piiper, Meyer, Worth & Willmer, 1977) are used paradigmatically, but the general approach is applicable to gills of teleost fish as well.

A qualitative model for analysis of gas exchange in fish gills is shown in Fig. 1. The entire process of gas exchange consists of three component processes; two are conductive and one is diffusive. They are ventilation (gill water flow), diffusion between water and blood and perfusion (gill blood flow). Their interaction in the counter-current model will be the subject of the core part of this chapter. Subsequently, complications arising from the curvilinearity of blood dissociation curves and diffusion resistance in water will be treated in some detail. Finally, other factors expected to complicate the application of the model to real fish gills will be briefly discussed.

The gas of primary interest is oxygen. The model and the equations derived

Fig. 1. Model for gas exchange of fish gills.

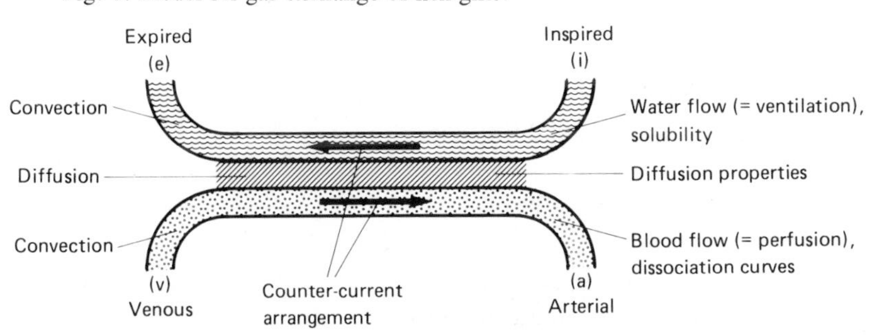

and applied in this chapter are, however, valid for the steady-state transfer of any gas between water and blood, but in particular for carbon dioxide. If gas transfer from water to blood is defined as positive, then carbon dioxide excreted in steady-state conditions exhibits a (formally) negative gas transfer rate, and hence all concentration and partial pressure differences for this gas assume a negative sign when positive for oxygen. However, when scaling partial pressure differences to their extreme values, the sign of the (dimensionless) ratio becomes independent of the sign of the transfer rate.

List of quantities and symbols To indicate the dimension, a consistent system of units is used. For b, h, l_0 and v, see Fig. 12.

C concentration (mmol l^{-1})
D diffusing capacity (transfer factor) ($\text{mmol min}^{-1}\ \text{Torr}^{-1}$)
d diffusion coefficient ($\text{dm}^2\ \text{min}^{-1}$)
F surface area (dm^2)
G conductance ($\text{mmol min}^{-1}\ \text{Torr}^{-1}$)
g thickness (dm)
K Krogh's diffusion constant ($\text{mmol dm}^{-1}\ \text{Torr}^{-1}\ \text{min}^{-1}$)
$\dot{M}$ transfer rate (mmol min^{-1})
P partial pressure (Torr)
$\dot{Q}$ perfusion (l min^{-1})
$\dot{V}$ ventilation (l min^{-1})
α capacitance coefficient (effective solubility) for water ($\text{mmol l}^{-1}\ \text{Torr}^{-1}$)
β capacitance coefficient (effective solubility) for blood ($\text{mmol l}^{-1}\ \text{Torr}^{-1}$)

Subscripts:

a arterial
b blood
e expired
i inspired
m membrane
v mixed venous
w water
w − b water–blood difference

The component processes

Ventilation

Gas transport by gill water flow or ventilation (sometimes termed irrigation) is schematically represented in Fig. 2. The transfer rate ($\dot{M}$) equals the product of the ventilation ($\dot{V}$) and the inspired–expired concentration difference ($C_i - C_e$), which can be expressed in terms of the inspired–expired partial pressure difference ($P_i - P_e$) when the capacitance coefficient (α) is introduced. Defining conductance, G, as transfer rate per partial pressure difference, one obtains the ventilation conductance $G_{vent} = \dot{V}\alpha$.

Since carbon dioxide is approximately 30 times more soluble in water than oxygen, the inspired–expired partial pressure difference for carbon dioxide must always be small, not exceeding a few Torr.

If P_{CO_2} is reduced to very low levels in carbonated water (seawater, 'hard' freshwater), bicarbonate is partially transformed into carbonate according to the buffering reaction

$$2\ HCO_3^- \rightarrow CO_2 + CO_3^{2-} + H_2O$$

The ensuing lowering of bicarbonate further steepens the carbon-dioxide dissociation curve of water, thereby increasing the capacitance coefficient (= effective solubility, dC/dP) above the physical solubility for carbon dioxide. The effective solubility ratio $\alpha_{CO_2}/\alpha_{O_2}$ is thus also increased.

Diffusion

Use is made of Fick's first diffusion law illustrated in Fig. 3. The diffusional transfer rate is proportional to the water–blood partial pressure difference P_{w-b}, and to the ratio effective area/effective thickness of the diffusion barrier. The diffusion conductance G_{diff}, defined as transfer rate per partial pressure difference, is termed diffusing capacity (D) in mammalian pulmonary physiology, and is alternatively called transfer factor. G_{diff} is proportional to Krogh's

Fig. 2. Ventilation (gill water flow). Model and transport equations. Right lower corner: concentration–partial pressure relations of oxygen and carbon dioxide in water, showing variable effective solubility or capacitance coefficient (α) for carbon dioxide in carbonated water.

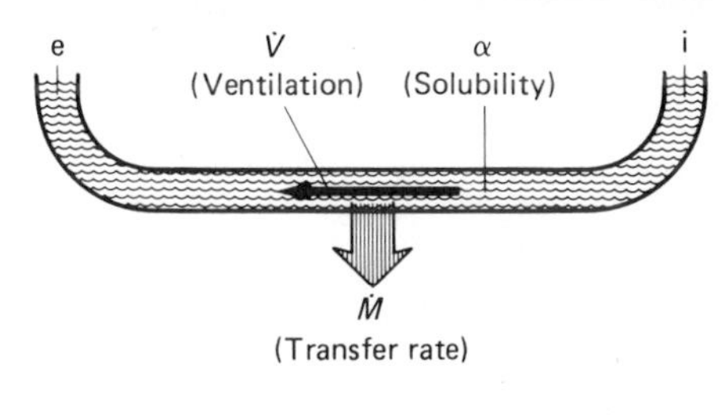

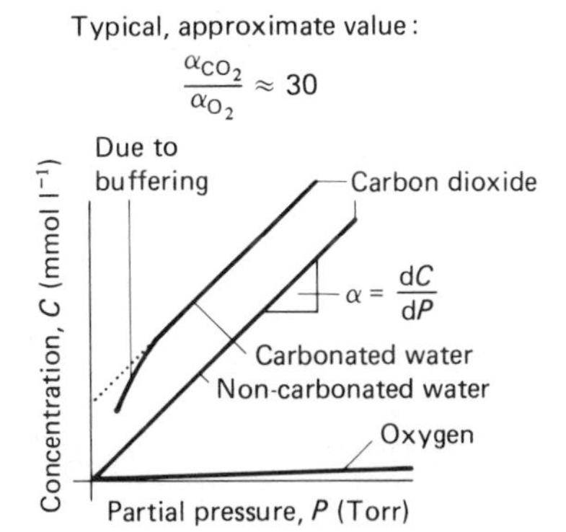

diffusion constant K, which is the product of diffusion coefficient (d) and solubility (α).

Due to the much higher α value, K in tissue and water is about 20 times higher for carbon dioxide than for oxygen. On this basis it is generally assumed that for carbon dioxide diffusion limitation plays only a minor role (see, however, below).

Perfusion

The equations for transport of gases by perfusion (gill blood flow) are formally identical with those for ventilatory transport (Fig. 4). (For simplicity,

Fig. 3. Diffusion through the water–blood barrier. Model and transport equations.

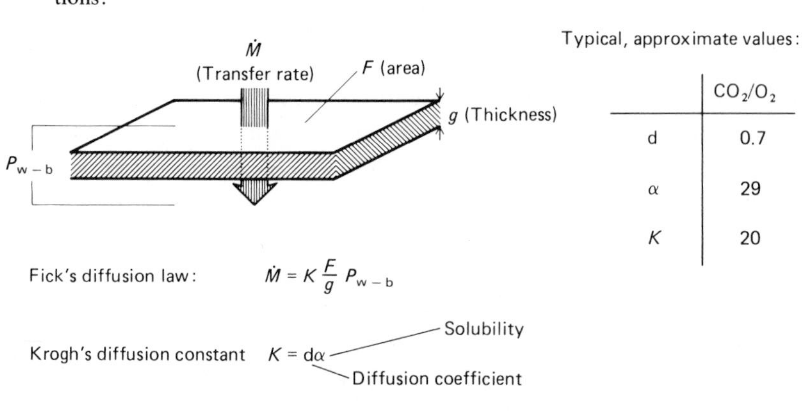

Fig. 4. Perfusion (gill blood flow). Model and transport equations. Right lower corner: concentration–partial pressure relations (dissociation curves) for oxygen and carbon dioxide in fish blood (schematic) to show the high β_{CO_2}/β_{O_2} ratio.

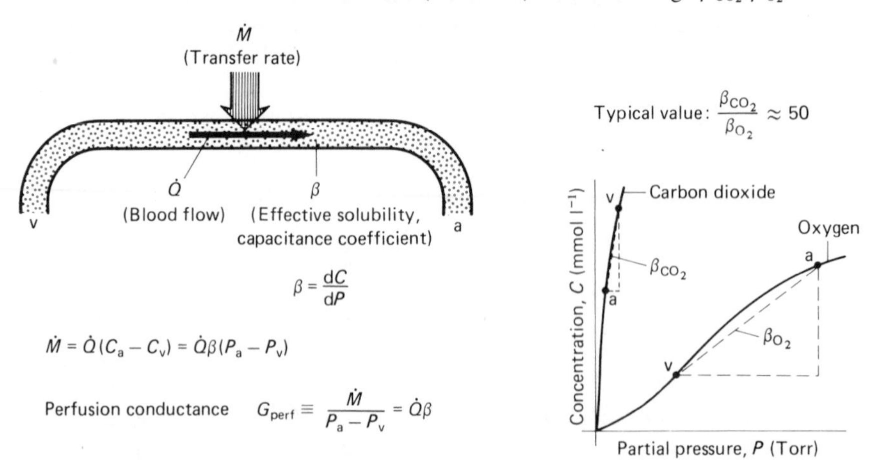

the subscript v replaces the more conventional subscript $\bar{v}$ for mixed venous blood; see the list of quantities and symbols.)

The blood dissociation curves for oxygen and carbon dioxide are curvilinear. Therefore the slope (β) of the line crossing the venous and arterial points of the dissociation curve, plotted as concentration (C) against partial pressure (P), varies with P. Hence the perfusion conductance, G_{perf}, which is defined as the transfer rate divided by the arterial–venous partial pressure difference, and is equal to the product of perfusion, $\dot{Q}$, and the effective solubility or capacitance coefficient for blood, β, also varies with the partial pressure range.

The carbon dioxide dissociation curve of blood is usually much steeper than the oxygen dissociation curve, a typical value for the β_{CO_2}/β_{O_2} ratio being 50. Therefore, the arterial–venous partial pressure differences for carbon dioxide are very small when compared with the corresponding partial pressure differences for oxygen.

Transfer of carbon dioxide versus oxygen

In most cases analysis of branchial exchange has been performed on the basis of oxygen measurements. This is mainly because in aquatic organisms oxygen availability is generally considered to be physiologically more important as the limiting factor than carbon-dioxide elimination. Secondly, as shown above, due to high G_{CO_2}/G_{O_2} ratios for ventilation, diffusion and perfusion, the P_{CO_2} differences are very small and therefore very difficult to determine experimentally with sufficient precision. In order to be able to understand carbon dioxide equilibria and acid–base balance relations, however, the quantitative analysis of carbon dioxide transport is a requirement.

Diffusion is generally considered as non-limiting in carbon dioxide transfer processes due to the high K_{CO_2} value (see above). However, the relative role of diffusion, as compared to ventilation and perfusion, in limiting transfer depends on the conductance ratios. In *Scyliorhinus stellaris,* the G_{diff}/G_{perf} and G_{diff}/G_{vent} ratios for carbon dioxide have been estimated to be of the same order of magnitude as those for oxygen (Piiper & Scheid, 1975), and they may easily be smaller, implying an even higher significance of diffusion limitation for carbon dioxide than for oxygen transfer.

Further specific problems in branchial carbon dioxide exchange will be mentioned later.

Counter-current model

It follows from the anatomical arrangement that in fish gills water and blood flow in opposite directions. Hence the branchial gas exchange takes place in a counter-current system with diffusive equilibration. The counter-current model is depicted in Fig. 5. It should be noted that the equations and the partial pressure profiles for oxygen presented in the figure are valid for constant α and β.

The partial pressure profiles in both water (w) and blood (b) are linear only when G_{vent}/G_{perf} equals unity. For lower or higher G_{vent}/G_{perf} ratios the partial pressure profiles of both water and blood are exponential curves. For $G_{vent}/G_{perf} > 1$, P_b approaches P_i exponentially, while for $G_{vent}/G_{perf} < 1$, P_w approaches P_v with an exponential profile.

In all the cases represented in Fig. 5, where L is the distance from the blood inflow end of the secondary lamellae and L_0 is the total length, there is an overlap of P values in blood and water, i.e. the size order of the partial pressures (decreasing order for oxygen, increasing for carbon dioxide) is P_i, P_a, P_e, P_v. The possibility of this overlap is a characteristic feature of the counter-current system as opposed to the co-current system (water and blood flowing in the same direction) in which the order is always P_i, P_e, P_a, P_v. This important distinction is visualised in Fig. 6. The partial pressure overlap corresponds to a high degree of gas exchange efficiency.

With increasing diffusive resistance (decreasing G_{diff}), however, the gas exchange efficiency of both systems is reduced and the partial pressure order becomes P_i, P_e, P_a, P_v for the counter-current system also. This is shown in

Fig. 5. Blood–water equilibration in counter-current model. Model, equations and partial pressure profiles for three selected cases (A, B and C).

(α = constant, β = constant)

$G_{vent} = \dot{V}\alpha$

$G_{diff} = D$

$G_{perf} = \dot{Q}\beta$

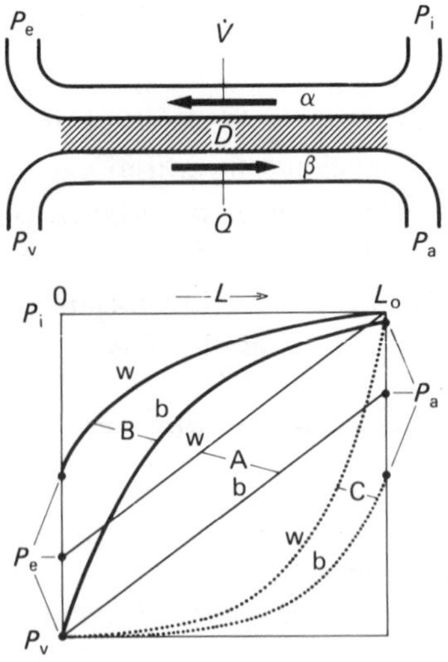

	A	B	C
G_{vent}/G_{perf}	1	2	0.5
G_{diff}/G_{perf}	3	3	5
G_{diff}/G_{vent}	3	1.5	10

$$\frac{P_i - P_e}{P_i - P_v} = \frac{1 - e^{-Z}}{X - e^{-Z}}$$

$$\frac{P_e - P_a}{P_i - P_v} = \frac{Xe^{-Z} - 1}{X - e^{-Z}}$$

$$\frac{P_a - P_v}{P_i - P_v} = \frac{X(1 - e^{-Z})}{X - e^{-Z}}$$

$$X = \frac{G_{vent}}{G_{perf}}$$

$$Z = \frac{G_{diff}}{G_{perf}} - \frac{G_{diff}}{G_{vent}}$$

$$\frac{P_w(L) - P_v}{P_i - P_v} = \frac{X - e^{-Z L/L_0}}{X - e^{-Z}} \qquad \frac{P_b(L) - P_v}{P_i - P_v} = \frac{X(1 - e^{-Z L/L_0})}{X - e^{-Z}}$$

Fig. 7 to occur at $G_{diff}/G_{perf} < 1$ (for $G_{vent}/G_{perf} = 1$). Furthermore, with $G_{diff}/G_{perf} < 0.5$ there is little difference in gas exchange efficiency between counter-current and co-current systems.

Although the counter-current system is the most efficient gas exchange system, it shows some unexpected features which may be detrimental to gas transport. The example C in Fig. 5 can be derived from the example A by doubling

Fig. 6. Comparison of counter-current and co-current models. The overlap of water and blood partial pressures present in the counter-current model (if G_{diff} is sufficiently high, see Fig. 7) does not occur in the co-current model.

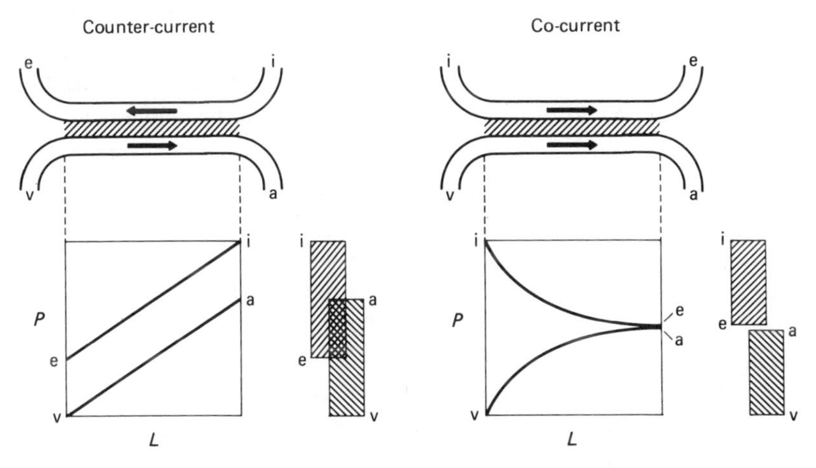

Fig. 7. Effect of diffusion limitation. The overlap of water and blood partial pressures (see Fig. 6) occurs only when G_{diff}, expressed here in terms of the G_{diff}/G_{perf} ratio, is high. With a low G_{diff}/G_{perf} ratio, the gas-exchange efficiency of the counter-current model approaches that of the co-current model.

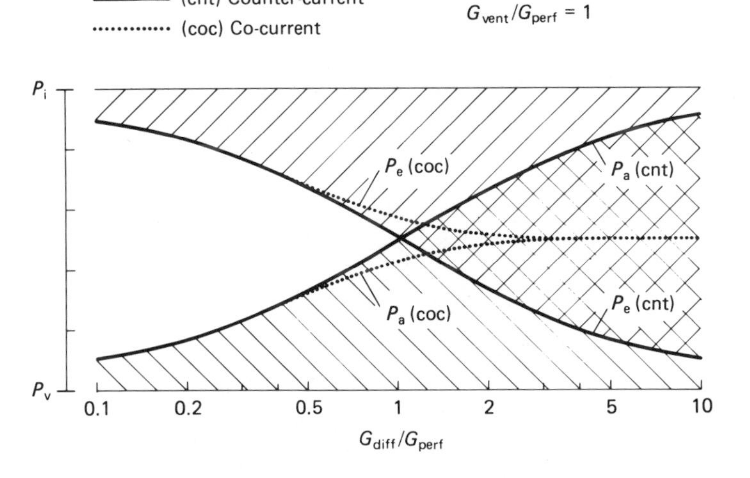

the perfusion (G_{perf}) at constant ventilation (G_{vent}). Remarkably, P_a approaches P_v, i.e. P_a falls for oxygen, even though G_{diff} is simultaneously increased 3.3 times.

In Fig. 8 the limiting case of $G_{diff} = \infty$ is analysed. At $G_{vent}/G_{perf} > 1$, perfusion (G_{perf}) limits gas exchange. An increase in G_{vent} in this case diminishes the inspired–expired partial pressure difference ($P_i - P_e$) to keep the transfer rate, G_{vent} ($P_i - P_e$) unaltered. At $G_{vent}/G_{perf} < 1$, gas exchange is ventilation-limited; an increase in perfusion (G_{perf}) reduces the arterial–venous partial pressure difference ($P_a - P_v$) in such a manner that the transfer rate remains constant. At $G_{vent}/G_{perf} = 1$, there remains no limitation, in that (small) changes in either G_{perf} or G_{vent} do not affect the transfer rate. In this case $P_a = P_i$ and $P_e = P_v$.

The fall of arterial blood P_{O_2} with increasing G_{perf} in a counter-current model is compared in Fig. 9 with the behaviour of the ventilated pool model of mammalian alveolar lungs. Upon an increase of G_{perf} in the counter-current model, at constant transfer rate, there is a large drop in arterial P_{O_2}, and a small rise in venous P_{O_2}. In the ventilated pool model, the major change is a rise in venous P_{O_2}, while arterial P_{O_2} remains practically constant. It should, however, be noted that P_a is above P_e in the counter-current model, whereas in the ventilated pool model, P_a remains below or at P_A (the partial pressure of oxygen in the alveolus). When G_{perf} approaches infinity, P_a approaches P_v and this value equilibrates to P_e or P_A in both models. With more extensive diffusion limitation, when $G_{diff} < G_{vent}$ and $G_{diff} < G_{perf}$, an increase in G_{perf} causes a rise in P_v and a fall in P_a by similar amounts in both models.

The basic counter-current model and the relations between the G and P values

Fig. 8. Effects of the G_{vent}/G_{perf} ratio in the counter-current model. No diffusion limitation, $G_{diff} = \infty$. $G_{vent}/G_{perf} < 1$, gas transfer is ventilation-limited; in the range $G_{vent}/G_{perf} > 1$, gas transfer is perfusion-limited.

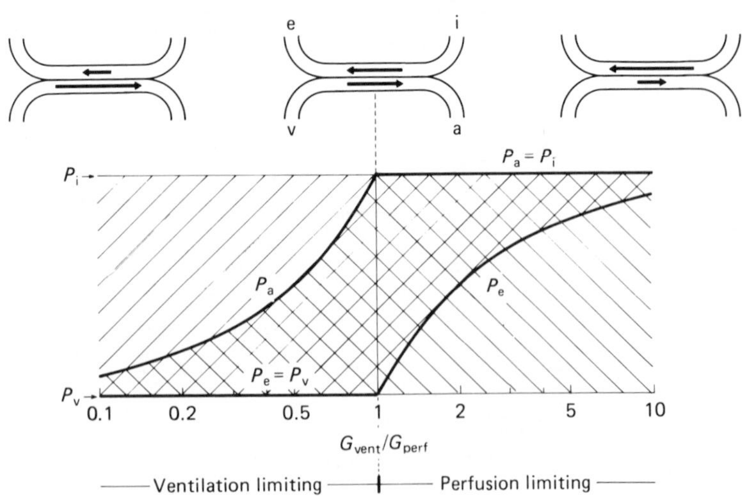

shown in Fig. 5 are precisely valid only under certain assumptions, the most important of which are the following.

(1) Water and blood dissociation curves are linear; α and β are constant.

(2) Water and blood flow are constant; $\dot{V}$ and $\dot{Q}$ are constant.

(3) All resistance to gas transfer is diffusive and is located in a barrier between water and blood.

(4) All the parallel secondary lamellar units are functionally identical, i.e. the system is homogeneous.

In the following sections the effects of deviations from these assumptions will be analysed.

Blood dissociation curves

Blood dissociation curves (equilibrium curves) for the respiratory gases oxygen and carbon dioxide are more or less markedly curvilinear. Hence β is

Fig. 9. Effects of an increase in perfusion conductance, G_{perf}, upon the P_{O_2} of arterial and mixed venous blood in both the counter-current (fish gills) and ventilated pool (mammalian lungs) models. P_{O_2} in inspired medium (water or air) and oxygen transfer rate ($\dot{M}$) constant. Models and P_{O_2} profiles. The decrease in $P_a - P_v$, produced by doubling of G_{perf}, is mainly due to an increase in P_v in the ventilated pool model, and to a decrease in P_a in the counter-current model.

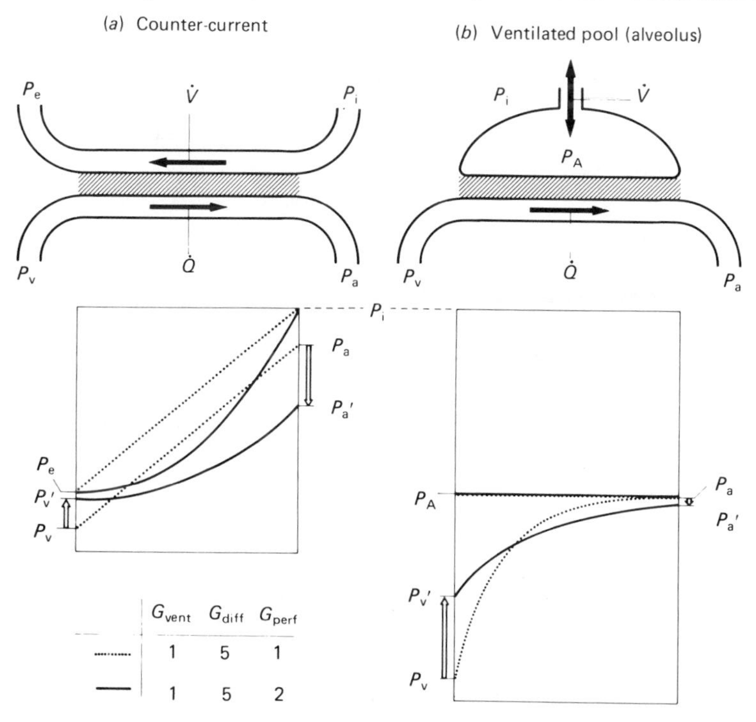

not constant, and the equations of Fig. 5, which have been derived by analytical integration of differential equations for constant α and β, are not precisely valid. However, it is always possible to subdivide the length of the counter-current model into sufficiently small elements (ΔL) in which the changes in partial pressures are small enough to render constant α and β valid assumptions. The transfer equations are then applied in each element

ventilation: $\Delta\dot{M} = \dot{V}\alpha\Delta P_{\mathrm{w}}$
diffusion: $\Delta\dot{M} = P_{\mathrm{w-b}}\Delta D$
perfusion: $\Delta\dot{M} = \dot{Q}\beta\,\Delta P_{\mathrm{b}}$

and the elemental changes are summed up over the total length L_0.

The determination of the diffusing capacity D ($=G_{\mathrm{diff}}$) from $\dot{M}$, P_{i}, P_{e}, P_{a} and P_{v} for oxygen by the numerical integration method is visualised in Fig. 10*b*. The underlying values are based on measurements obtained in the elasmobranch fish *Scyliorhinus stellaris* (Piiper & Baumgarten-Schumann, 1968*b;* Piiper *et al.*, 1977). The total length of the counter-current model is subdivided into a number of elements (here ten). The oxygen uptake, $\Delta\dot{M}$, and hence the arterial–venous oxygen concentration difference and the inspired–expired oxygen concentration difference, are set equal for all elements. The diffusing capacity of each element (ΔD), corresponding to the length of the element, is calculated as oxygen uptake per water–blood $P_{\mathrm{O_2}}$ difference, $P_{\mathrm{w-b}}$, of the element (visualised for element 1 at the venous–expiratory end of the model). Finally, the total D is obtained as the sum of ΔD.

Fig. 10. Determination of diffusing capacity, $D = \dot{M}/\bar{P}_{\mathrm{w-b}}$. Models, equations and procedure. For explanations, see text.

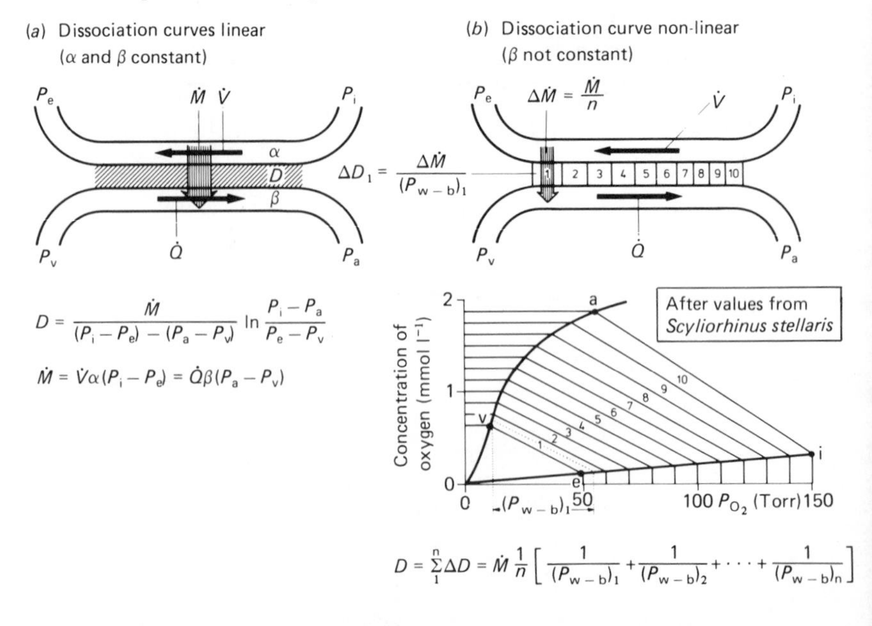

In Fig. 10*a* the analytical solution for constant α and β is presented. Note that for the case $G_{vent}/G_{perf}=1$ $(P_i-P_e=P_a-P_v)$ the solution is: $D=\dot{M}/(P_i-P_e)=\dot{M}/(P_a-P_v)$.

Piiper *et al.* (1977) reported the following ratio of D_{O_2} values from the same experimental data obtained from the dogfish *Scyliorhinus stellaris*

$$D_{O_2}\text{ (oxygen dissociation curve)}/D_{O_2}\text{ (}\beta_{O_2}\text{constant)} = 0.90$$

This result can easily be understood considering that the curvature of the oxygen dissociation curve provides a larger water–blood P_{O_2} difference, thus enhancing diffusive oxygen uptake. Therefore a smaller D value is required for the same oxygen uptake.

The carbon dioxide diffusing capacity has been determined using the same procedure (Piiper & Baumgarten-Schumann, 1968*b*).

Diffusion resistance of interlamellar water

Since the width of the interlamellar space is considerable (ranging from 20 to 100 μm), whereas that of the water–blood tissue barrier is usually much less (ranging from 0.2 to 10 μm) (cf. Piiper, 1971), a sizable part of the diffusion resistance to gas exchange is expected to reside in interlamellar water. As gases dissolved in interlamellar water are transported both by diffusion and by convection, a comprehensive analysis is difficult.

In Fig. 11*a* the range of the partial pressure within a cross-section of an interlamellar space is represented by the width of the stippled band. The profile cal-

Fig. 11. Models for the study of diffusion resistance in gill water. Models, partial pressure profiles and equations for apparent diffusing capacity, D_{app}. For explanations, see text.

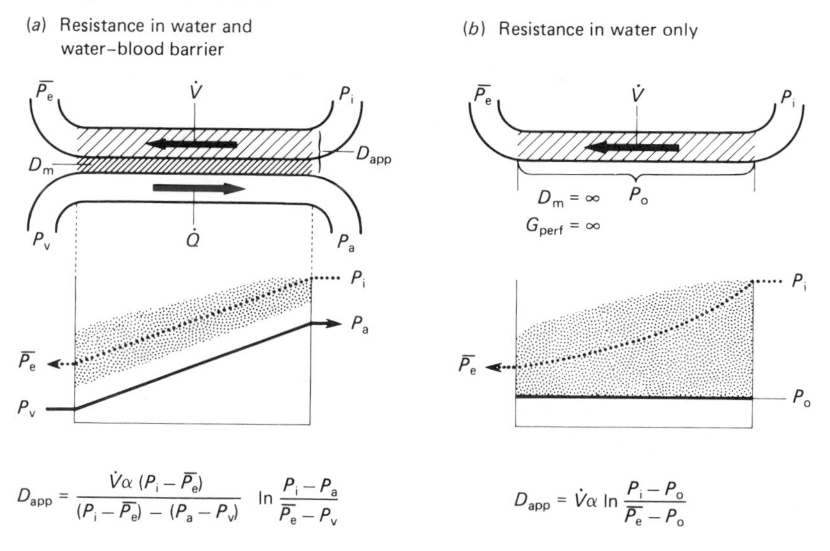

culated upon the assumption that the whole resistance is located in a barrier between water and blood (in accordance with the model of Fig. 5) is represented by the dotted line. For this assumption an apparent G_{diff} (D_{app}) can be calculated using the equation which is derived from the equations of Fig. 5 (see also Fig. 10).

In order to investigate the diffusion resistance of interlamellar water separately, the model shown in Fig. 11*a* is simplified by setting $D_{\mathrm{m}} = \infty$ and $G_{\mathrm{perf}} = \infty$. In the resulting model (Fig. 11*b*), gas transfer between water and the lateral wall, with partial pressure constant at P_{o} (i.e. the wall acting as an infinite sink or source), is governed by water flow and by diffusion in water. Again, the model with diffusion resistance in the water–blood barrier (Fig. 5) can be used to calculate the profile of partial pressure in water (simple exponential approach to P_{o}) and to obtain a value for D_{app}. The ratio of the D_{app} values for the models shown in Figs. 11*a* and *b* should yield a rough estimate of the fraction of diffusional resistance which is located in interlamellar water (see below).

Gas equilibration for the model shown in Fig. 11*b* has previously been examined in theory (Scheid & Piiper, 1971). The geometry of the model is shown in

Fig. 12. Model for diffusive equilibration in interlamellar space. Model, $P_{\mathrm{O_2}}$ profiles along the inscribed x, y and z axes, differential equation for gas transport by diffusion and convection, and equation for combined parabolic and hyperbolic water velocity distribution.

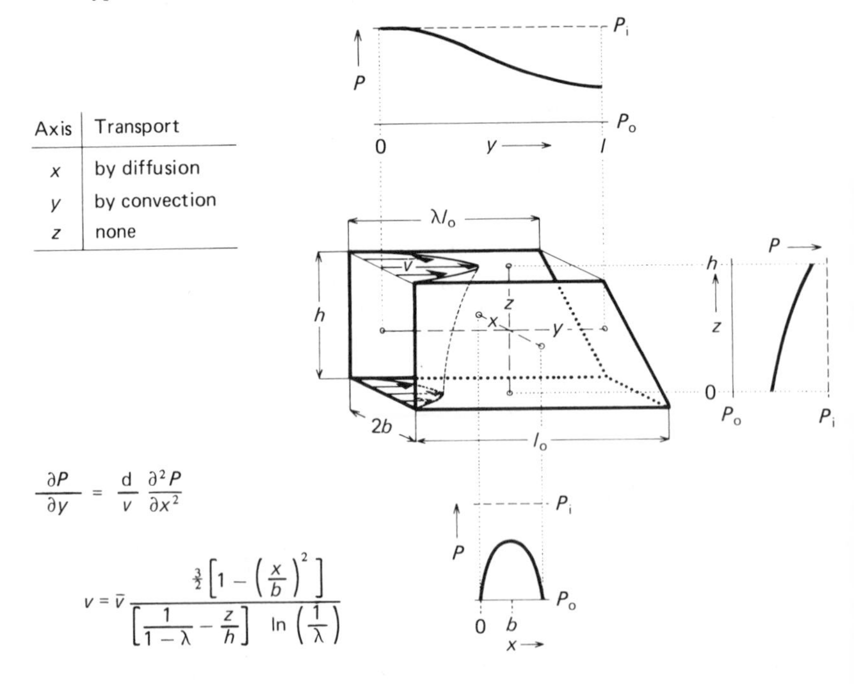

Fig. 12. The water flow is assumed to be laminar, and there is a vertical velocity gradient due to the tapering of the lateral walls (= secondary lamellae) from base to top. The gas (oxygen) is assumed to be transported in the x-direction by diffusion only, and in the y-direction by convection only. The partial differential equation for diffusion and convection can be solved numerically by finite step methods using a computer. The gas partial pressures show gradients in all three space coordinates (x, y and z). For a given P_i and P_o, the partial pressure in mixed expired water, $\bar{P}_e$, turns out to be a function of the width of the space (b), the base length (l_o), the diffusion coefficient (d), the mean water velocity ($\bar{v}$) and the taper (λ). (h is the height of the secondary lamella.) In Fig. 13, the equilibration efficiency, defined as $(P_i - \bar{P}_e)/(P_i - P_o)$, is shown to decrease with increasing ϕ $(= b^2\bar{v})/(l_o \mathrm{d})$ and with decreasing λ.

Calculations with physiological and morphometric measurements taken on the dogfish *Scyliorhinus stellaris* have shown that about half of the resistance to diffusion is located in water, and half in the water–blood barrier (Scheid & Piiper, 1976). This result, however, is subject to criticism because mathematically the diffusion resistance of water and that of the water–blood barrier are not additive. Therefore, we recently began to investigate a more comprehensive model, corresponding to that shown in Fig. 11*a*, in which diffusion in the barrier and diffusion in water are represented by two separate conductance terms, and G_{perf} is finite. The model is shown in Fig. 14.

According to our preliminary results the overall effective G_{diff} (Fig. 11*a*) is somewhat higher than the G_{diff} value obtained for the model in Fig. 11*b* and

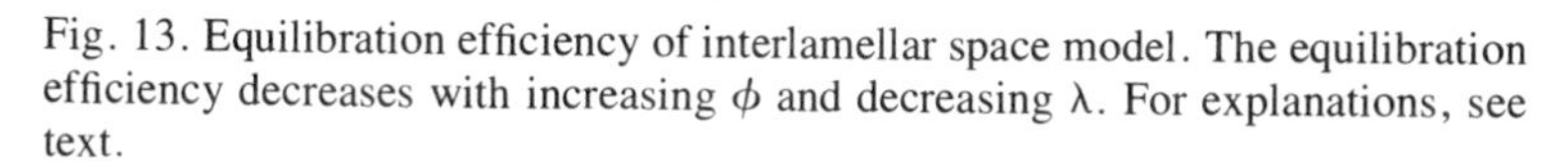

Fig. 13. Equilibration efficiency of interlamellar space model. The equilibration efficiency decreases with increasing ϕ and decreasing λ. For explanations, see text.

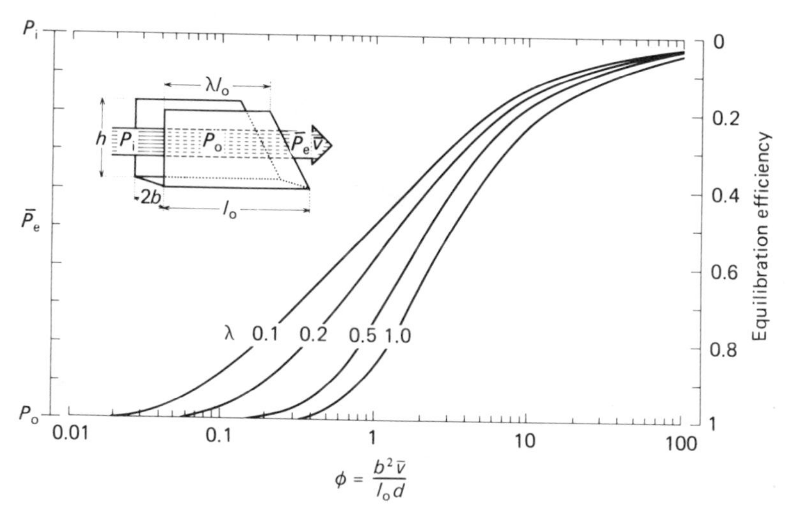

combined with the value obtained from barrier morphometry ($G_{diff(m)}$). When applied to the previous analysis performed by Scheid & Piiper (1976) this means that the experimentally observed oxygen exchange is less efficient than the oxygen exchange predicted on the basis of morphometric and gill water-flow measurements. This discrepancy may well derive from inaccuracies in the numerous data used for the calculations, but it is also possible, and even probable, that other factors, which are briefly discussed in the following section, come into play.

Further complicating factors

Besides diffusion resistance in interlamellar water and curvilinearity of blood dissociation curves, there are several further complicating factors which produce deviations from the behaviour of the simple counter-current equilibration model shown in Fig. 5.

Limiting processes within blood

Part of the total diffusion resistance is located in the blood (plasma and red blood cells). Also, the reaction of oxygen with haemoglobin may play an important limiting role in oxygen uptake, particularly at low temperatures. Both diffusion and reaction resistances may be included in an equilibration capacity of blood for oxygen.

Even more complex are the processes known to be involved in carbon dioxide exchange. In mammalian blood the carbonic-anhydrase-accelerated hydration and dehydration of carbon dioxide and carbonic acid respectively as well as the

Fig. 14. Comprehensive model for analysis of gas exchange in fish gills. Limitation by diffusion in interlamellar water (w) and in water–blood barrier (m) and by perfusion. Model and equations for conductances.

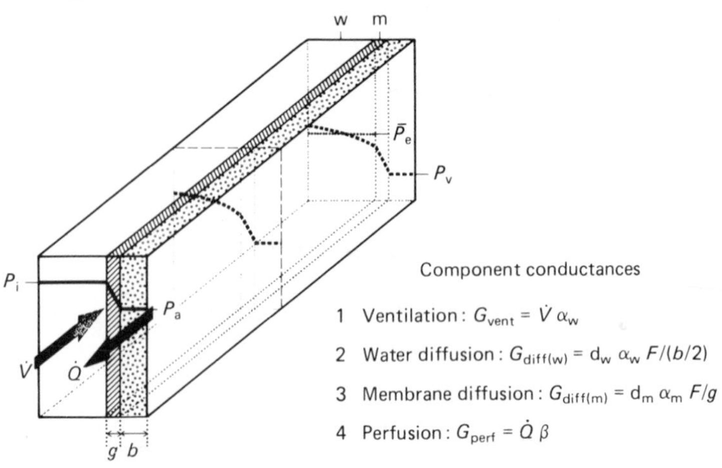

chloride/bicarbonate exchange between red cells and plasma (Hamburger shift) have been suspected as processes which limit pulmonary and tissue carbon dioxide exchange. Cameron and Polhemus (1974) have produced models for the limiting role of blood carbonic anhydrase in carbon dioxide equilibration in the gills of rainbow trout. In freshwater teleost blood there seem to exist particular limitations to erythrocyte/plasma transfer of bicarbonate, and the excretion of carbon dioxide has been conceived to be a process controlled and limited by epithelial enzymatic and electrolyte permeability functions of the gill rather than by ventilation and diffusion of carbon dioxide (Haswell, Randall & Perry, 1980).

All such additional resistances reduce gas exchange efficiency. When, in the analysis according to the simple model of Fig. 5, they are not taken into account specifically, they will cause a reduction of the apparent D value for carbon dioxide. It should be noted that in the marine elasmobranch *Scyliorhinus stellaris* the carbon dioxide equilibration was found to be remarkably efficient, leaving little space for a limiting role of other processes besides the diffusion of carbon dioxide (Piiper & Baumgarten-Schumann, 1968*b*).

Cyclic variations in water and blood flow

Gill water flow and gill perfusion vary within the ventilatory and cardiac cycles respectively. The effect of pulsatile flow on gas exchange is to decrease the efficiency particularly when the extent of diffusion limitations is small.

A synchronisation of heart beat with respiration in such a manner that maximum water and blood flow occur simultaneously would be helpful, particularly at low ventilatory and cardiac frequencies. Although the breathing and heart rate frequencies in fish are often very similar, such coordination seems to be present only to a very limited extent (Taylor & Butler, 1971; Hughes, 1972).

Parallel inhomogeneities

Fish gills comprise a great number of secondary lamellae and interlamellar spaces as units of gas exchange. If the units are not equal with respect to anatomical dimensions, ventilation or blood flow, parallel inhomogeneity will arise, reducing the gas exchange efficiency in a manner similar to the corresponding well-known effects of unequal distribution of ventilation and perfusion in mammalian lungs. According to the anatomical structure, several orders of maldistribution are expected; (1) between gill arches or gill slits, (2) between filaments and (3) between secondary lamellae or interlamellar spaces.

Shunting of water and blood

A shunt is formally an extreme form of unequal distribution, meaning the presence of gas exchange units with finite G_{vent}, but no G_{perf} or G_{diff} (water shunt), and units with finite G_{perf}, but no G_{vent} or G_{diff} (blood shunt). A water shunt could easily arise from the movement apart of gill arches or gill filaments, for example during great breathing efforts. A blood shunt may arise from blockage of water passages or perfusion of vessels in the base of the secondary lamella.

It should be noted that blood flow via anastomotic vessels arising from the efferent and afferent branchial arteries and leading to the venous system would not cause a shunt in terms of gas exchange.

Implications

All the above-considered deviations of the real situation in fish gills from the simplified model tend to diminish the overall gas exchange efficiency. When the definitions of G_{vent} and G_{perf} are retained, all these factors formally lead to a reduction of G_{diff} $(=D)$ as calculated on the basis of Fig. 5, i.e. neglecting the disturbing inhomogeneity effects. At present no adequate method is available for estimating the effects of these factors on gas exchange in fish gills.

References

Baumgarten-Schumann, D. & Piiper, J. (1968). Gas exchange in the gills of resting unanaesthetized dogfish (*Scyliorhinus stellaris*). *Respiration Physiology,* **5,** 317–25.

Cameron, J. N. & Polhemus, J. A. (1974). Theory of CO_2 exchange in trout gills. *Journal of Experimental Biology,* **60,** 183–94.

Haswell, M. S., Randall, D. J. & Perry, S. F. (1980). Fish gills carbonic anhydrase: acid–base regulation or salt transport? *American Journal of Physiology,* **238,** R240–5.

Hughes, G. M. (1972). The relationship between cardiac and respiratory rhythms in the dogfish, *Scyliorhinus canicula* L. *Journal of Experimental Biology,* **57,** 415–34.

Piiper, J. (1971). Gill surface area: fishes. In *Biological Handbooks: Respiration and Circulation,* ed. P. L. Altman & D. S. Dittmer, pp. 119–21. Bethesda, Maryland: Federation of American Societies for Experimental Biology.

Piiper, J. & Baumgarten-Schumann, D. (1968*a*). Transport of O_2 and CO_2 by water and blood in gas exchange of the dogfish (*Scyliorhinus stellaris*). *Respiration Physiology,* **5,** 326–37.

Piiper, J. & Baumgarten-Schumann, D. (1968*b*). Effectiveness of O_2 and CO_2 exchange in the gills of the dogfish (*Scyliorhinus stellaris*). *Respiration Physiology,* **5,** 338–49.

Piiper, J., Meyer, M., Worth, H. & Willmer, H. (1977). Respiration and circulation during swimming activity in the dogfish *Scyliorhinus stellaris*. *Respiration Physiology,* **30,** 221–39.

Piiper, J. & Scheid, P. (1972). Maximum gas transfer efficacy of model for fish gills, avian lungs and mammalian lungs. *Respiration Physiology,* **14,** 115–24.

Piiper, J. & Scheid, P. (1975). Gas transport efficacy of gills, lungs and skin: theory and experimental data. *Respiration Physiology,* **23,** 209–21.

Piiper, J. & Schumann, D. (1967). Efficiency of O_2 exchange in the gills of the dogfish, *Scyliorhinus stellaris*. *Respiration Physiology,* **2,**135–48.

Scheid, P. & Piiper, J. (1971). Theoretical analysis of respiratory gas equilibration in water passing through fish gills. *Respiration Physiology,* **13,** 305–18.

Scheid, P. & Piiper, J. (1976). Quantitative functional analysis of branchial gas transfer: theory and application to *Scyliorhinus stellaris* (Elasmobranchii). In *Respiration of Amphibious Vertebrates,* ed. G. M. Hughes, pp. 17–38. London and New York: Academic Press.

Taylor, E. W. & Butler, P. J. (1971). Some observations on the relationship between heart beat and respiratory movements in the dogfish (*Scyliorhinus canicula* L.). *Comparative Biochemistry and Physiology,* **39A,** 297–305.

LEONARD B. KIRSCHNER

Physical basis of solute and water transfer across gills

Introduction

Gills may be regarded as regions of the body surface of aquatic animals that are specialised to maximise the transfer of oxygen and carbon dioxide. These gases move by diffusion, and the structure and behaviour of the gill and its microcirculation illustrate adaptive responses of a biological unit to the demands of physical law; in this case to the parameters of the diffusion equations. The adaptations that facilitate respiratory gas movement engender, in many animals, new problems in maintaining ionic and osmotic steady states in their body fluids. The purpose of this chapter is to review the physical bases for dissipative* flows. The treatment will be neither exhaustive nor rigorous. Instead, emphasis has been placed on what the equations tell us about gills, and how they have been used (and misused) in studying branchial function. A number of general reviews of solute and solvent movement have been published, as well as some on specialised aspects. The thermodynamic basis of molecular transfer was described by Katchalsky & Curran (1965), and this framework was used to describe and analyse active sodium transport across epithelia (Caplan & Essig, 1977). Koch (1970) provided a good didactic treatment of some useful solutions of the diffusion equation. Water movement in living systems was reviewed by House (1974), who provided both a clear treatment of theory and a comprehensive overview of the information available.

The gill can be considered as an interface, of thickness x, that separates two compartments each of which is filled with an aqueous solution; one the external medium and the other the extracellular fluids within the animal. The situation is depicted for solutions of a single solute s in Fig. 1. Net movement of s through the interface will develop if a driving force is applied to the molecules in a direction normal to the plane of the membrane. The necessary force is usually a gradient of potential energy across the membrane, but it may also be due to movement of the centre of mass of the solution in the membrane; i.e. the driving force is applied to the entire solution rather than to s alone (as in filtration). For

*Such a flow is called dissipative because it reduces the potential energy stored in the gradient unless energy is expended to maintain a constant gradient while flow is occurring.

the present we will deal only with the former, called by Koch (1970) 'interfusion', and by Mauro (1957) 'molecular drift'. In principle, any modality of energy, for example a thermal gradient, might drive the solute flux, and the symbol *G,* the Gibbs' free energy, is used in Fig. 1 to suggest this. However, in biological systems it appears that gradients of chemical potential and electrical potential suffice to account for most passive (diffusive) solute fluxes. Provided that the system is not far from equilibrium, the laws relating flows with applied forces appear to be linear. If solute molecules act independently of each other, and we can neglect any bulk flow of solvent, net solute flux across the interface is described by the Nernst–Planck diffusion equation

$$J = -A\mathrm{D}[\mathrm{d}c/\mathrm{d}x + (\mathrm{zF}/\mathrm{R}T)\, c\, \mathrm{d}\psi/\mathrm{d}x] \quad (1)$$

where J is the net flux of solute (mol s^{-1}), A is the area of membrane over which diffusion takes place, c is the concentration†, ψ is the electrical potential, z is the valence of s, F the Faraday constant, R the gas constant, and T the absolute

Fig. 1. A two-compartment system is shown with aqueous solutions of a solute (s) separated by a membrane (M) permeable to the solute. The potential energy of the solute, expressed by the symbol for its Gibbs' free energy, is higher in the outer (G_s^o) than in the inner solution (G_s^i). The bulk solutions are well mixed, but a stagnant (unstirred) layer is apposed to each surface of the membrane.

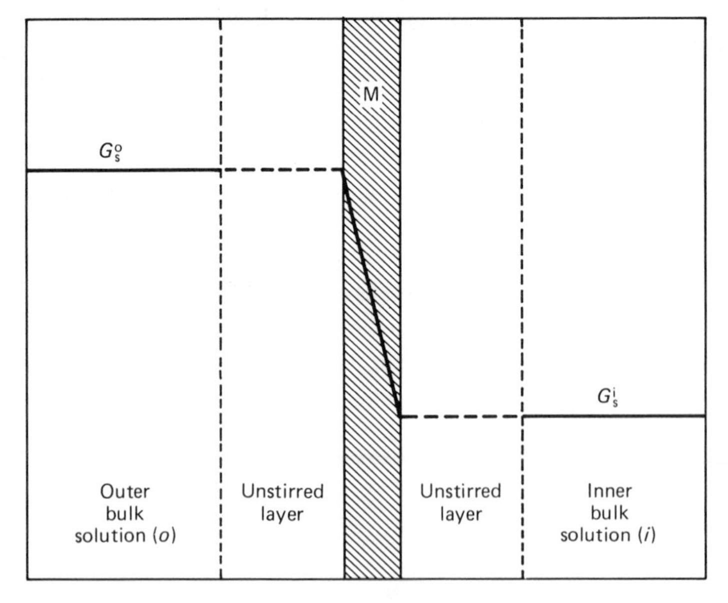

†In fact, the chemical activity, γc, should be used, but we rarely know the activity coefficient in biological work, and concentration is a generally accepted approximation.

temperature. The proportionality constant D is called the diffusion coefficient ($cm^2 s^{-1}$). This cryptic statement contains the information which animals need to develop adaptations for coping with dissipative flows, maximising those that are useful and minimising those likely to be harmful. However, eqn (1) is not generally useful; that is, there is no general analytical solution relating the measurable quantities, fluxes, concentrations and potential difference across the surface. However, it is possible to specify a condition or set of conditions that make it possible to integrate the equation, so that the solution expresses the flux or other variable in terms of measurable quantities. Such 'boundary conditions' are simply behaviours of the system in which diffusion is occurring. They may be demonstrable (the solute is a non-electrolyte, for example) or assumed (the electrical gradient is linear through the membrane); either may yield explicit, useful solutions of eqn (1), and examples of both will follow. For the present it suffices to emphasise that a particular solution is useful only if the system behaves in the specified manner, and the onus is on the investigator to show that it does. Unfortunately, this burden is not always shouldered.

Gas exchange across the body surface

Oxygen transport across the body surface appears to be diffusive, hence net influx should conform with eqn (1). Fortunately, the system animal–environment–oxygen has properties that constitute useful boundary conditions. In the first place, molecular oxygen is uncharged; $z = 0$, and the entire electrical term in eqn (1) drops out. Therefore, we can write

$$J^{O_2}_{net} = -AD[dc/dx] \qquad (2)$$

which is Fick's diffusion equation for non-electrolytes. Another simplification is possible. The external environment can be considered as an infinite reservoir of oxygen, and the amount transferred to an animal has no effect on its concentration (c_{out} is constant). In addition, because the branchial blood is flowing past the exchange surface, entry of oxygen does not change its concentration *at any point* along the exchange surface. Therefore, c_{plasma} (c_{in}) at that point is also constant. Actually, as blood flows parallel to the exchange surface its oxygen concentration rises between afferent and efferent ends, and this might appear to complicate the picture. However, water, on the outside of the exchange surface, flows in the opposite direction (counter-current) to the blood, and its oxygen concentration falls between the points of entry and of exit. The situation is depicted in Fig. 2, and it can be seen that a constant concentration difference between medium and blood is maintained, at least approximately. In consequence, $dc/dx = \delta c/x$, and eqn (2) can be rewritten

$$J^{O_2}_{net} = -AD[c^{in} - c^{out}]/x \qquad (3)$$

where c_{in} and c_{out} are accessible to measurement. The parameters of eqn (3), the exchange area (A), thickness (x), and diffusion coefficient (D) will determine the oxygen flux for any concentration difference. The value of D is unknown; its value may be fixed by the lipoprotein structure of cell membranes. But the importance of the other two is apparent in the microanatomy of the gill and its microcirculation. Clearly, the shorter the diffusion distance the more rapid will be diffusive oxygen transfer. The epithelium of the respiratory lamellae, which separates branchial blood from the external medium, is only 2–10 μm thick (Hughes & Wright, 1970). By way of contrast, the single cell layer in the vertebrate nephron, another transport epithelium, is about 30 μm thick. Since D is unknown and a value for x may be uncertain, their ratio, $D/x = p$, called the permeability coefficient (cm s^{-1}), is usually used to express the conductance (or permittivity) of the gill for oxygen.

Eqn (3) also shows that transfer is faster when the area for exchange (A) is large. Measurement of the respiratory exchange area in fish (Hughes, 1966) and crustaceans (Gray, 1957) shows that this is generally large by comparison with the general body surface. Moreover, among fish at least, external branchial surface area tends to be larger in active animals with a high oxygen requirement (Hughes, 1966). However, the area may also be limited from the inside, by

Fig. 2. The physical situation favouring oxygen transfer is shown. Blood and water flow in opposite directions (counter-current) along internal and external surfaces of the branchial epithelium. The numbers suggest concentrations, usually expressed as partial pressures. At any point along the surface a concentration gradient favours oxygen diffusion from water to blood. Although gas tensions change along the plane of the epithelium the counter-current arrangement should maintain the gradient approximately constant. The oxygen flux is shown by the arrow normal to the epithelial surface.

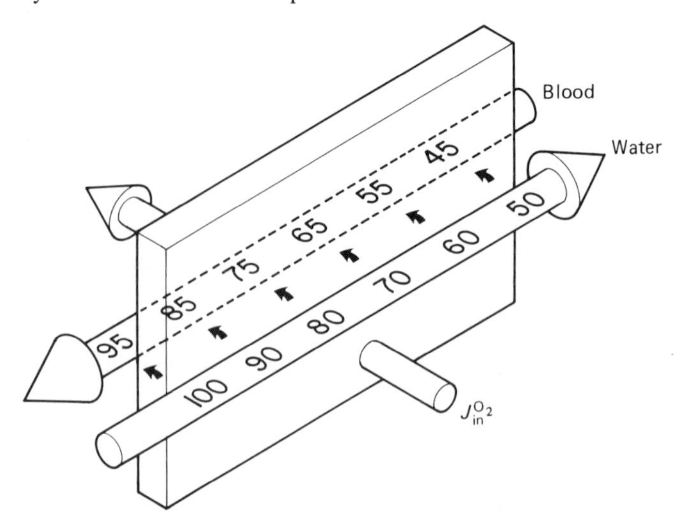

altering the number of gill filaments perfused by the blood. This cannot be measured, but it, and not the external surface, is probably the limiting area under most conditions. For this reason, the quantity AD/x is a lumped parameter that must be considered a practical permeability. This is done, in fact, by respiratory physiologists who call it the 'transfer factor'. Note that a change in the area perfused by blood will cause the diffusional flux to change even if the gradient remains constant and with D, x and external area fixed. Changes in the transfer factor, observed when fish become active (e.g. Randall, Holton & Stevens, 1967), probably have such a basis.

Carbon dioxide excretion in aquatic animals has been assumed to occur as in the mammalian lung where the driving force is simply a concentration difference between plasma and medium carbon dioxide. If this is correct then the general principles, developed above for oxygen transfer, would apply. However, the model has been challenged recently by Haswell, Randall & Perry (1980). In an alternative model, plasma bicarbonate moves into the branchial cells where it is dehydrated. The carbon dioxide produced then diffuses out into the medium. If the first step is rate-limiting, the kinetics would be controlled by the movement of an anion rather than uncharged carbon dioxide. Eqn (3) would be inappropriate, because the boundary condition $z = 0$ would not apply at the limiting step.

Ion movement across the body surface

In many marine animals the most abundant ions in both environment and extracellular fluids are sodium and chloride, and to a first approximation the two solutions are isoionic. Since there are no gradients, diffusional movements present no problem. However, some marine animals maintain blood sodium chloride levels well below the ambient level (this is also true for potassium, magnesium, calcium and sulphate ions), and all freshwater animals maintain extracellular ionic concentrations far above the ambient. These animals are not in diffusional equilibrium, and dissipative fluxes threaten them with salt loading or loss. This would pose no problem if gills were impermeable to ions, but the structural adaptations permitting high rates of gas transfer are apparently incompatible with ion impermeability, and diffusive movement must be considered. Ionic diffusion is also governed by eqn (1), and it is necessary to seek boundary conditions that generate useful solutions. The nature of the information sought, and sometimes the available tools, determine what constitutes a useful solution.

One question that sometimes arises concerns quantitative evaluation of gill permeability to an ion (p), or, if the area is unknown, a practical permeability analogous to the transfer factor described above. Following the lead of neurophysiologists, we can *assume* that the electrical field across the gill is constant (i.e. the potential gradient is linear). With this boundary condition, and if the net flux is constant, eqn (1) becomes, after rearranging

$$dc/dx + [zF/RT]c\delta\psi/x = -J/AD$$

which has a solution

$$J_{net} = -Ap(zF/RT)V_g\ [c_{out} - c_{in}\ \exp\ (zF/RT)\ V_g]/1 - \exp(zF/RT)\ V_g \quad (4)$$

where V_g ($=\psi_{in} - \psi_{out}$) is the potential difference across the gill and the other terms have the same meaning as before. The concentrations and voltage are accessible, and if J_{net} can be measured, p (or Ap) can be calculated. This is straightforward in principle, but measurement of a net flux is often impractical. Instead an isotope is usually used to measure the total 'unidirectional' flux from one compartment to the other (i.e. J_{in} or J_{out}). Since $J_{net} = J_{in} - J_{out}$ it is plausible that eqn (4) can be decomposed to give

$$J_{in} = -Ap(zF/RT)V_g\ c_{out}/1 - \exp(zF/RT)\ V_g \quad (4a)$$

$$J_{out} = Ap(zF/RT)V_g\ c_{in}\ \exp(zF/RT)\ V_g/1 - \exp(zF/RT)\ V_g \quad (4b)$$

This conjecture was shown to be true by House (1963) and Smith (1969). Eqn (4a) was used recently to show that the gill of a marine fish was an order of magnitude more permeable to sodium ions than to chloride ions (Kirschner & Howe, 1981).

One of the most frequent questions that arises in this context is whether an observed flux is purely diffusional or contains a component due to *active transport*. Transport has a number of distinguishing characteristics, but in many cases the only practical criterion is whether the observed behaviour can be described satisfactorily by the diffusion equation. Briefly, the test is to measure the components of the gradient, concentrations and voltage, and from them to predict the diffusional behaviour. The actual behaviour is then measured and compared with the calculated one. If the two correspond then the behaviour is purely diffusional; if they differ appreciably we suppose that the difference is due to active transport. The first problem is to find boundary conditions that generate a useful solution of eqn (1). The solution expressed by eqn (4) can be used, but uncertainty about the values of A and p, and the possibility that they might change during an experimental procedure, renders conclusions ambiguous; this approach must be used cautiously. A popular method is to abolish the electrochemical gradient by making concentrations equal on both sides of the epithelium and, using an external circuit, to abolish any potential difference between the two chambers (a technique called voltage-clamping or short-circuiting the preparation). Under these conditions $dc/dx = 0$, $d\psi/dx = 0$, and hence $J_{net} = 0$; diffusion is abolished. A net flux under these conditions must be due to transport. The technique was developed by Ussing (Ussing & Zerahn, 1951) and has been used, largely in his laboratory, to establish most of the conceptual basis for active ion transport. Unfortunately, the complex morphology of gills does not lend itself to voltage clamping; the procedure has never been used on these organs.

The procedures most commonly used in work on intact animals involve the use of the flux–ratio equation and the Nernst equation, both of them being solutions of eqn (1). The first was derived by Ussing (1949). In this method the unidirectional fluxes (J_{in} and J_{out}), rather than the net flux, are considered (note that $J_{net} = J_{in} - J_{out}$). If the diffusion equation is solved for each unidirectional flux, the ratio of the two is a simple function of concentrations and voltage.

$$J_{in}/J_{out} = (c_{out}/c_{in}) \exp[-(zF/RT)\, V_g] \quad (5)$$

The flux–ratio equation has the virtue of being independent of both permeability and exchange area. The only assumption made in its derivation is that J_{net} is constant in time, and no special experimental conditions are required to satisfy boundary conditions. Practically, one need only measure concentrations and voltage to calculate the ratio for an ion moving by diffusion alone, then measure the two fluxes with isotopes. If measured and calculated ratios agree, the fluxes are purely diffusive. Otherwise transport is involved.

The other criterion requires the special constraint that the ion is in diffusional equilibrium. This is true when $J_{net} = 0$ ($J_{in} = J_{out}$). If this condition is applied to eqn (1) (or to eqns (3) or (4)) the solution is

$$V_g = (RT/zF) \ln c_{out}/c_{in} \quad (6)$$

That is, providing that an ion is at diffusional equilibrium, its distribution can be predicted if we know V_g. In addition, if one of the concentrations is changed abruptly, V_g should change in a predictable fashion. Eqn (6) is called the Nernst equation, and the value of V_g at which the specified distribution is at equilibrium is called the equilibrium potential, V_{eq}. The test for transport is especially simple in this situation. The concentrations are measured and V_{eq} calculated. Then V_g is measured; if $V_g = V_{eq}$ diffusion alone can account for the distribution found. No fluxes need be measured.

Water movement across gills

The flow–force relations for water have the same general shape as for solute; that is, water movement is proportional to a gradient of its potential. However, the mole fraction of water in aqueous solution is high and its structure so complex that the molecules do not act independently of each other. As a result, the chemical potential is not even approximated by its concentration for most purposes. However, in one type of measurement, the self-diffusion of water, the treatment given for solutes is adequate. Here, a small quantity of labelled water (e.g. [^{2}H] water or [^{3}H] water) is added to a volume of ordinary water. The tracer concentration is low and can be treated like any uncharged solute; its movement is described by eqn (2). As before, D and x are usually unknown, and the diffusional permeability to water is usually expressed in terms of the permeability coefficient, $p_d = D/x$. Published values of fluxes are usually normalised for area based on measurements of the gill's external surface, whereas the internal area might be the limiting surface, as pointed out earlier. The parameter shown is more complex than is usually assumed and has the properties of the transfer factor used for gas exchange.

For other types of water flow experiments the potential gradient is expressed differently. As is well known, if an interface, as in Fig. 1, is water-permeable, a difference in hydrostatic pressure (ΔP) will engender water flow. The volume of water which flows is proportional to the pressure difference: $J_v \propto \Delta P$. In addition, the presence of a solute will lower the chemical potential of water in a solution. We can determine the effect of the solute empirically as follows (it can also be derived from thermodynamic principles). Let a solute-impermeable membrane separate pure water from an aqueous solution of solute *s*. If the membrane is water-permeable, flow will occur from pure water into the solution. This is osmotic flow (osmosis) and the gradient causing it is suggested in Fig. 3. This flow can be abolished if a hydrostatic pressure is applied to the solution compartment. The chemical activity of water in the presence of a solute can be expressed as the pressure necessary to abolish water flow under the conditions specified above. This is the osmotic pressure, π, of the solution. Note that the osmotic pressure of pure water is zero, and that osmotic flow occurs from lower to higher pressure. This may be intuitively uncomfortable but presents no practical difficulties. Volume flow is proportional to the osmotic pressure differences: $J_v \propto \Delta\pi$.

The molecular basis of water flow in the presence of hydrostatic and osmotic gradients is not well understood. It is clear that the independent movement of molecules is not quantitatively important; for example, it accounted for much less than 1% of the total flow when a pressure difference was applied across a

Fig. 3. The test system used in the text to define the osmotic pressure (π). The inner compartment (to the right) always contains pure water, and the outer compartment contains the test solution. The two are separated by a mechanically rigid membrane that is permeable to water but impermeable to solute. Pressure is applied to the outer compartment to abolish net water flow.

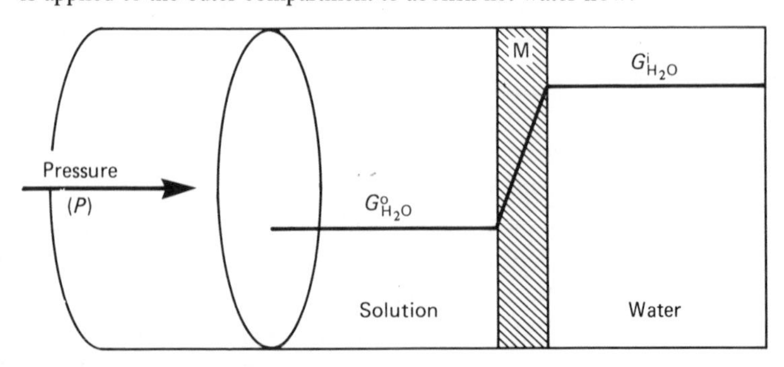

collodion membrane (Mauro, 1957). But whatever the mechanism, the flow–force relations have been well characterised, and a number of important points emerge. In the first place, equal osmotic and hydrostatic pressures generate equal water flows; therefore, the hydraulic and osmotic conductivities are the same (Durbin, 1960), and

$$J_v = A\mathrm{L_p}(\Delta P - \Delta\pi) \tag{7}$$

where J_v is the volume flux of water (cm^3 s^{-1}), and $\mathrm{L_p}$ is the hydraulic conductivity (cm s^{-1} atm^{-1}). Secondly, osmotic pressure increases with solute concentration, and it can be shown that the relation is described by the gas law, $\pi = c_s \mathrm{R}T$, where c_s = mol cm^{-3}, R = cm^3 atm^{-1} deg^{-1} mol^{-1}, and T is the absolute temperature. Moreover, the species and characteristics of the solute play no special role; as long as the membrane is impermeable, one mole of any solute, even a single ion, exerts an osmotic pressure of 22.4 atm in a litre of solution under standard conditions. It follows that a mixture of solutes exerts the same osmotic pressure as a single solute of equal concentration providing that our test membrane (Fig. 3) is impermeable to all of them. Therefore, eqn (7) can be written

$$J_v = A\mathrm{L_p}(\Delta P - \mathrm{R}T\Delta\Sigma c) \tag{7a}$$

where $\Delta\Sigma c$ is total solute concentration difference. Finally, the definition of osmotic pressure used above requires that the interface is completely impermeable to the solute. Most biological membranes and epithelia are not completely solute-impermeable, and it is of practical importance to be able to characterise the osmotic behaviour of a leaky system. This amounts to the introduction of a third membrane coefficient (the first two characterise water and solute permeabilities) to describe the interaction of the two diffusive flows. The 'reflection coefficient', σ, was introduced by Staverman for this purpose. It is usually defined in terms of the coefficients of the thermodynamic equations linking flows and forces, but can also be defined by a simple experiment. If the membrane used to define osmotic pressure is replaced by one that is solute-permeable we find that the pressure needed to abolish J_v is reduced; the leakier the membrane the less pressure is required. In general, for a single solute

$$\Delta P_{J_v=0} = \sigma\Delta\pi = \sigma\mathrm{R}T\Delta c_s \tag{8}$$

The parameter is obviously dimensionless. If the membrane is impermeable to solute, $\sigma = 1$; if solute and solvent conductances are the same, $\sigma = 0$, and for most purposes this encompasses the range of values. It is clear that if $\mathrm{L_p}$ is to be estimated from osmotic flow data the value(s) of σ for any solute(s) present must be known, otherwise the estimate is a lumped parameter, $\mathrm{L_p}/\sigma$. It would be dimensionally correct but would overestimate the correct value. This is obviously a problem for work on gills, which are known to be permeable to the major solute present (sodium chloride), and which have recently been shown to be permeable to non-electrolytes as large as inulin (mol. wt = 5000) (Kirschner, 1980). Unfortunately, no estimates of σ are available, and the correction is never applied; probably the correction is small ($\sigma \sim 1$) in ionic regulators, since a very leaky gill would be maladaptive. However, salt flow is very rapid in isoionic marine invertebrates (cf. Fig. 1 in Kirschner, 1979), and L_p might be grossly overestimated by an osmotic measurement in these animals.

Thus, either of two parameters can be used to characterise water flow across an interface; the diffusional permeability, p_d(cm s^{-1}) and the hydraulic coefficient, L_p(cm s^{-1} atm^{-1}). Since they have different units they cannot be compared directly. However, if we express the water flow in eqn 7 in the same units (mol s^{-1}) as in eqn (6)

$$J_v/\bar{V} = J_D = (L_p RT/\bar{V})\ (\Delta P/RT - \Delta c_s) \tag{9}$$

where $\bar{V}$ is the partial molar volume of water (cm^3 mol^{-1}). The quantity $L_p RT/\bar{V}$ is called the osmotic permeability, p_{os}, and has the same dimensions as p_d which makes a comparison of permeability coefficients determined in tracer and osmotic experiments possible. If the same mechanism underlies flow in both measurements the coefficients should be equal. Many of the values published for gills are based on relatively crude measurements of J_v (e.g. drinking rates), but the results suggest that $p_{os} > p_d$ in most cases. This is a very general phenomenon, having been observed in artificial membranes, cells and epithelia (cf. Tables 4.4, 5.5 and 9.5 in House, 1974). It supports the surmise that volume flow with a pressure or osmotic gradient is only partly diffusive, and that another mechanism is also involved. However, the question is complicated by a methodological problem that might be especially severe in an organ as complex as a gill.

The effect of unstirred layers on tracer measurements

Even with a simple planar membrane bathed by well-stirred solutions there is a static layer of fluid apposed to each side; that is, a layer in which there is no convective flow and hence no stirring. Such a layer is shown on each side of the membrane in Fig. 1. Movement in these layers is limited by diffusion, and under some conditions may lead to inaccurate estimates of the membrane coefficients. The reason is not hard to appreciate qualitatively. In Fig. 1 a membrane with low permeability to s separates two well-stirred bulk solutions at concentrations $c_s^o > c_s^i$. The solute will diffuse from compartment o, through an unstirred layer and across the membrane into the other stagnant layer, and then into solution i. If membrane permeability is sufficiently low this step will be rate-limiting, and diffusion in the unstirred layers will suffice to maintain the concentration in these layers equal to that in the bulk solutions; the entire gradient develops across the membrane, as in Fig. 1. In Fig. 4 the membrane is very permeable to *s*, and diffusion in the unstirred layers is not fast compared with the rate of transfer across the membrane. In this case gradients develop in the unstirred layers and the driving force across the membrane is less than $c_s^o - c_s^i$, the concentrations in the bulk phases. But the latter is what we measure (as for use in eqn (3)), and the real permeability will be underestimated.

The problem can be treated quantitatively as follows. Each unstirred layer can be considered as a membrane with a permeability coefficient $p = D/\delta$, where D is the diffusion coefficient of *s* in water and δ is the thickness of the layer. The resistance to diffusion in each phase is $1/p$, hence the total resistance across the two stagnant layers and the membrane, $1/p_T$, is given by

$$1/p_T = 1/p_o + 1/p_m + 1/p_i = \delta_o/D + 1/p_m + \delta_i/D \quad (10)$$

where p_o and p_i are the permeabilities of the unstirred layers and p_m that of the membrane. It is the latter that we usually want to estimate, but p_T is the value

Fig. 4. The effect of unstirred layers on the concentration profile of a solute to which the membrane is very permeable ($p_s > D/\delta$, where δ is the unstirred layer thickness). The concentration difference used to calculate p is $c_s^o - c_s^i$. The actual difference is $c_s^{mo} - c_s^{mi}$ (where c_s^{mo} and c_s^{mi} are the concentrations of *s* at the outer and inner surfaces of the membrane respectively) which is smaller. As a result, p_s is underestimated.

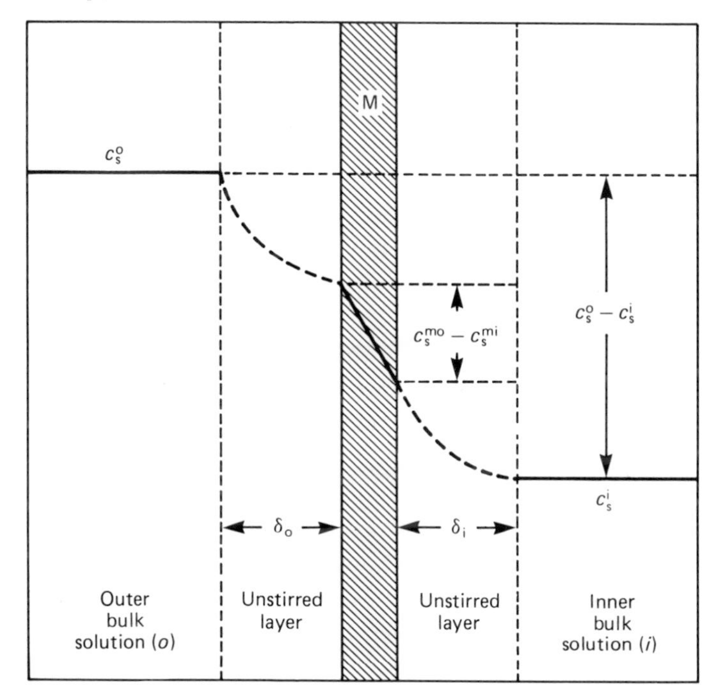

obtained when bulk concentrations are used. Eqn (10) can be rearranged to show that

$$p_T = ([D/(\delta_o + \delta_i)]p_m)/([D/(\delta_o + \delta_i)] + p_m) \tag{11}$$

When $p_m >> D/(\delta_o + \delta_i)$ the apparent permeability approaches $D/(\delta_o + \delta_i)$ and the estimate is determined by diffusion through the unstirred regions. On the other hand when $p_m << D/(\delta_o + \delta_i)$, $p_T \sim p_m$. It is obvious that the larger the value of δ, the layer thickness, the more likely it is to become limiting.

The severity of the unstirred layer problem in gills has not been assessed, but it may well distort values for the permeability coefficients for oxygen and for tracer water diffusion, because these molecules are so permeant in the branchial epithelium. Part of the apparent discrepancy between p_d and p_{os} for water might result from an underestimate of the former.

Coupling of flows

The movement of one molecular species, either by diffusion or by active transport, often creates a force that acts on a second species, causing the latter to move. A simple example involves movement of solute from one compartment to the other in a system originally at osmotic equilibrium. Imagine that the inner compartment in Fig. 1 contains a second solute, y, to which the membrane is impermeable. If $(s + y)_{inner} = s_{outer}$ there is no osmotic gradient, but the diffusion gradient of s still exists. Flow of s will dilute the outer and concentrate the inner solution, and this will cause osmotic water flow in the same direction as the solute vector. Water flow is said to be coupled to solute diffusion. Similarly, transport or diffusion of a single ion will change the potential difference between two compartments and engender the movement of other ions. Among epithelia, coupled flows are prominent in the small intestine and proximal renal tubule. It is one of the virtues of non-equilibrium thermodynamics that it provides a quantitative description of such coupling. However, there is no evidence that coupling of dissipative flows is important in gills, and the subject will not be pursued further here.

There is one type of flux-coupling which *is* of interest in connection with the role of the gill in hydromineral balance in animals capable of ionic regulation. Because sodium chloride is the major osmotically active solute in both the external environment and the body fluids, a gradient of salt favouring diffusion in one direction is also an osmotic gradient promoting water movement in the other. A 'typical' freshwater animal might have a body fluid sodium chloride concentration of about 100 mmol l^{-1} (200 mOsmol) with an environmental concentration <1 mmol l^{-1}. There is not only a gradient favouring loss of sodium chloride, but also one favouring inflow of water. It is likely that σ_{NaCl} is near unity in the gill, so that the effect of salt diffusion on osmosis should be small. Nevertheless, there remains a biological problem for the animal and a methodological problem for the investigator. Osmotic water uptake tends to increase the animal's volume. The animal's response is to excrete a volume of urine equal to the osmotic flow. This serves to maintain a stable volume. However, no animal is able to produce a solute-free urine, and urine sodium chloride concentrations are typically 10–100 times higher than those in the medium. This means that there are two routes of salt loss in these animals; by diffusion across the gills, described by eqn (1), and through the urine ($=c_{NaCl}J_v$ where c_{NaCl} is the urine concentration and J_v is the urine flow). As a consequence, in coping with the osmotic problem by excreting urine, the animal exacerbates the problem of maintaining blood sodium chloride concentrations. In many animals solute loss in the urine is greater than that achieved by diffusion across the gill, and the requirement for active inward transport to maintain an ionic steady state is correspondingly greater. A similar problem faces marine hypoionic regulators (for a more complete discussion see Kirschner, 1979).

The methodological problem governs which solution of eqn (1) to use in analysing flow-force behaviour in intact animals. This can be illustrated by example. A trout in seawater is in an ionic steady state, therefore, $J_{in}=J_{out}$ for sodium chloride. In approaching the question of whether these ions are actively transported across the gill one has to decide whether to use eqn (5) or (6). Since $J_{net}=0$, the latter appears to be appropriate, but this is not the case. More than half of the total influx occurs in the gut after ingestion of the medium, and J_{in}/J_{out} *across the gill* is 0.3–0.5, not 1.0; eqn (5) must be used. This is an instance of the general requirement mentioned earlier: the experimental situation must satisfy any boundary conditions imposed by a particular analytical solution of the diffusion equation.

References

Caplan, S. R. & Essig, A. (1977). A thermodynamic treatment of active sodium transport. In *Current Topics in Membranes and Transport,* vol. 9, ed. F. Bronner & A. Kleinzeller. New York: Academic Press.

Durbin, R. P. (1960). Osmotic flow of water across permeable cellulose membranes. *Journal of General Physiology,* **44,** 315–26.

Gray, I. E. (1957). A comparative study of the gill area of crabs. *Biological Bulletin of the Marine Biology Laboratories, Woods Hole,* **112,** 34–42.

Haswell, M. S., Randall, D. J. & Perry, S. F. (1980). Fish gill carbonic anhydrase: acid–base regulation or salt transport? *American Journal of Physiology,* **238,** R240–5.

House, C. R. (1963). Osmotic regulation in the brackish water teleost, *Blennius pholis. Journal of Experimental Biology,* **40,** 87–104.

House, C. R. (1974). *Water Transport in Cells and Tissues.* London: Edward Arnold.

Hughes, G. M. (1966). The dimensions of fish gills in relation to their function. *Journal of Experimental Biology,* **45,** 177–95.

Hughes, G. M. & Wright, D. E. (1970). A comparative study of the ultrastructure of the water–blood pathway in the secondary lamellae of teleost and elasmobranch fishes – benthic forms. *Zeitschrift für Zellforschung und Mikroskopische Anatomie,* **104,** 478–93.

Katchalsky, A. & Curran, P. R. (1965). *Nonequilibrium Thermodynamics in Biophysics.* Cambridge, Massachusetts: Harvard Press.

Kirschner, L. B. (1979). Control mechanisms in crustaceans and fishes. In *Mechanisms of Osmoregulation in Animals,* ed. R. Gilles. London: John Wiley and Sons.

Kirschner, L. B. (1980). Uses and limitations of inulin and mannitol for monitoring gill permeability changes in rainbow trout. *Journal of Experimental Biology,* **85,** 203–11.

Kirschner, L. B. & Howe, D. (1981). Exchange diffusion, active transport and diffusional components of transbranchial Na^+ and Cl^- fluxes. *American Journal of Physiology,* in press.

Koch, A. R. (1970). Transport equations and criteria for active transport. *American Zoologist,* **10,** 331–46.

Mauro, A. (1957). Nature of solvent transfer in osmosis. *Science,* **126,** 232–3.

Randall, D. J., Holton, G. F. & Stevens, E. D. (1967). The exchange of oxygen and carbon dioxide across the gills of rainbow trout. *Journal of Experimental Biology,* **46,** 339–48.

Smith, P. G. (1969). The ionic relations of *Artemia salina* (L.) II. Fluxes of sodium, chloride and water. *Journal of Experimental Biology,* **51,** 739–57.

Ussing, H. H. (1949). The distinction by means of tracers between active transport and diffusion. *Acta Physiologica Scandinavica,* **19,** 43–56.

Ussing, H. H. & Zerahn, K. (1951). Active transport of sodium as the source of electric current in the short-circuited, isolated frog skin. *Acta Physiologica Scandinavica,* **23,** 110–27.

CHARLOTTE P. MANGUM

The functions of gills in several groups of invertebrate animals

In 1941 August Krogh published an essay entitled *The Comparative Physiology of Respiratory Mechanisms* which has served both as a primer and as a source of stimulation for several generations of respiratory physiologists. In it he made the point (p. 31) that 'the normal way for aquatic animals to supplement the respiration through the general body surface is by appendages which are in the zoological literature usually called gills when no other major function can be assigned to them'. He also noted that 'gills' perform functions other than gas exchange and, indeed, that many of the structures zoologists call gills may have little or no respiratory function. Although Krogh was not explicit on the nature of these additional functions in 1941, it is clear from his other writings that he had in mind the processes which are responsible for salt and water balance with the medium, and for the excretion of ammonia (Krogh, 1939). In 1981 we should also note that feeding is often the primary function of some gills, which was suggested by the work of Hazelhoff (1938) and later demonstrated more quantitatively by Jørgensen (1952). Finally, we should turn the point around and note that structures designated as feeding organs by zoologists, such as the tentacles of sea anemones, may also function as gills in gas exchange (Sassaman & Mangum, 1973).

In the present essay I shall discuss the recently investigated functions of gills in three groups of animals. In one case a structure called a gill does not perform all the functions that we expect, at least in a higher group of animals. In the second example, the physiological processes usually performed by gills take place elsewhere for at least part of the day, and therefore the role of the gill is not obvious. And in the third group, an amazing diversity of structures, only some of which are called gills, perform several branchial functions.

The function of the *Limulus* gill

The book gills of the chelicerate arthropod *Limulus polyphemus* are borne on five pairs of modified appendages which are located on the ventral surface of the opisthosoma. Each one consists of about 150 rounded lamellae (Fig. 1*a*) which are about 3–4 cm wide and 2 cm tall in a 1 kg animal. The lamella is built as a chitinous envelope lined with a thin epithelial layer of cells and held together by crescent-shaped blocks of fibres (Fig. 1). The descriptions of blood flow through the gill differ on the point of whether the blood enters on the lateral side and exits on the medial side, or the reverse, but they agree that it flows around the periphery in one direction or the other (Lochhead, 1950; Shuster, 1978). The periphery is separated from a thicker central region by a thin line that resembles chitin, although its actual composition is not yet known (Fig. 1*a*). A section through the less dense periphery indicates that the cell lining is at most about 3.2 μm thick, and the blood–water diffusion distance is no greater than about 5.6 μm. A section through the central region suggests that its density is due to the thickening of only one face of the envelope, where the cell lining can be up to 10 μm thick (Fig. 1*c*).

Muscular movements of the opisthosomal appendages ventilate the gills sequentially in a rippling motion, like flipping through the pages of a book. These movements are also believed to aid in swimming, which occurs with the animal upside down. When the appendages move forward, the lamellae are brought apart anteriorly and ventrally, thus opening the pages of the book; when the flaps push backward and generate the thrust, the pages of the book are closed.

As suggested by its structure, the *Limulus* gill is a very fragile structure. The animal does not survive puncture of the gill epithelium by a very small hypodermic needle, probably because there is too little tissue nearby to repair the damage. When the animal encounters a lethal salinity, death appears to result from an inability to control the volume of the extracellular fluid as this lack of control causes ballooning of the lamellae, rupture of the very fragile epithelium and, finally, uncontrolled haemorrhaging.

We have suggested that the *Limulus* gill is primarily a site of gas exchange, and that at least two other processes, which are largely carried out by the gill in other aquatic arthropods, are performed elsewhere in *Limulus* (Towle, Mangum, Johnson & Mauro, 1982). These functions are osmoregulation, which I shall discuss only briefly because it is the province of other contributors to this volume, and acid–base regulation, which is traditionally treated by respiratory physiologists.

Osmoregulation

The evidence that the gill is not an important site of active ion transport is based to some extent on its structure, which simply does not contain nearly as much tissue as the osmoregulatory regions of the gill in other arthropods (Copeland & Fitzjarrell, 1968). However, studies of the ultrastructure of the gill are still under way, and we may learn that a very thin epithelium can have the cytological features that we associate with ion transport. More convincingly, we can find very little evidence of sodium ion transport, in the form of activity of the

Fig. 1. The gill of *Limulus polyphemus*. (*a*) Whole mount of an excised lamella fixed in Bouin's solution and stained with toluidine blue. P, thin peripheral region; C, thicker central area. Magnification: ×3.2. (*b*) Longitudinal section through thin peripheral region. B, blood space; BE, branchial epithelium; F fibrous connectives. Magnification: ×349. (*c*) Longitudinal section through thick central region. T, thickening of epithelium. Magnification: ×465.

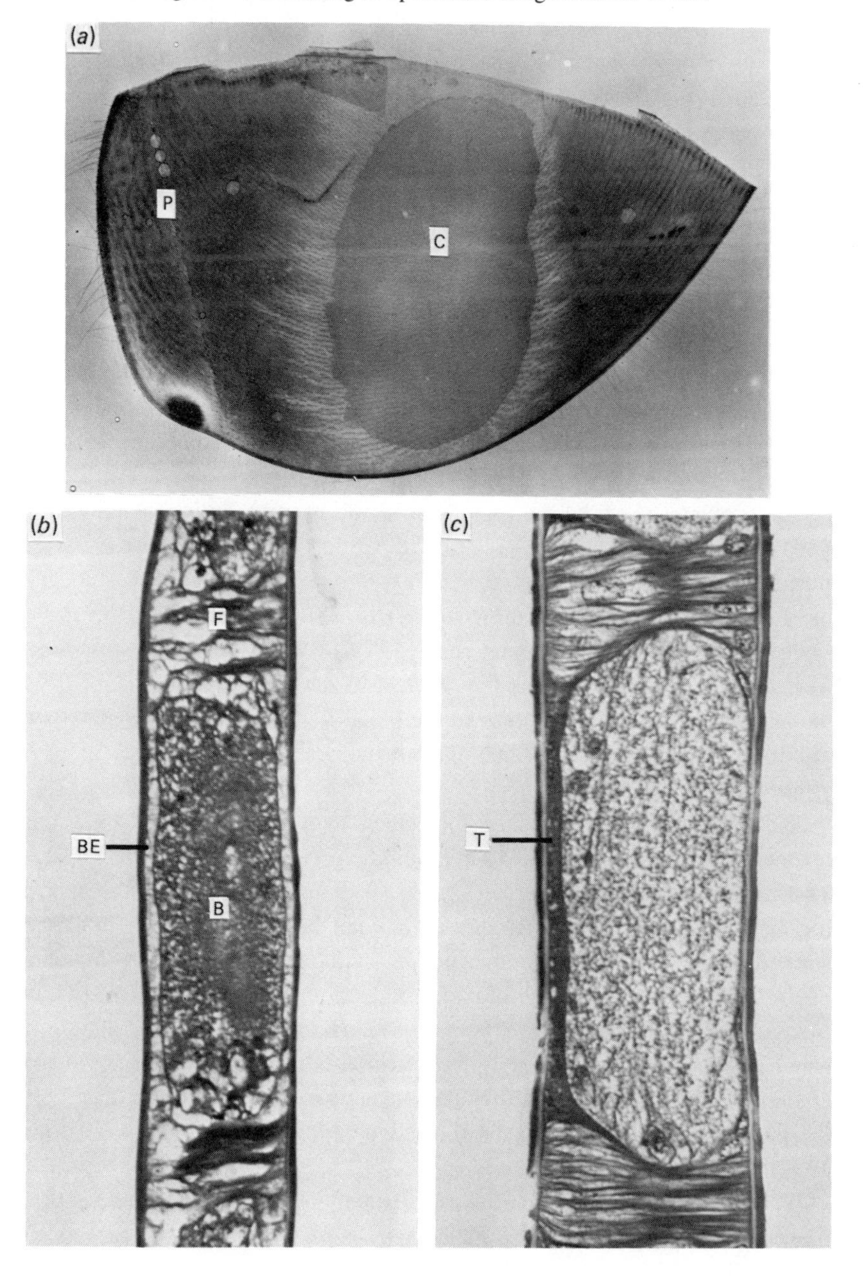

sodium- and potassium-dependent ATPases, at the cellular level (Towle *et al.*, 1982). In no other example known to us does a sodium-chloride absorbing epithelium lack appreciable activity of these enzymes, an activity which is enhanced when net sodium chloride uptake is activated (Towle *et al.*, 1982).

Moreover, there is a viable alternative hypothesis, viz. that, unlike other euryhaline arthropods, *Limulus* utilises its excretory organ (the coxal gland) as a major site of osmoregulation. The supporting evidence includes the ultrastructure of the coxal gland, which has many of the requisite features of a transporting epithelium; the high activities of the sodium- and potassium-dependent ATPases, which increase further at low salinity, and the ionic and osmotic composition of the blood and the urine. Unlike euryhaline crustaceans, the urine of an osmoregulating *Limulus* is always low in sodium chloride and total osmolality (Towle *et al.*, 1982).

Acid–base regulation

I mention the control of salt and water balance between the gill and the medium because the process is often coupled to acid–base regulation, at least at a physiological level (Evans, 1975; Mangum, 1976*a*; also other chapters in this volume).

The problem of acid–base regulation in the haemocyanin-containing arthropods is especially interesting because the pH dependence of their oxygen transport systems is so great. The constant expressing the change in haemocyanin–oxygen affinity with pH ($\Delta \log P_{50}/\Delta pH$) is -0.92 (pH 7.2–7.8) in the crab *Libinia emarginata,* for example, and 0.53 (pH 6.91–7.50) in *Limulus* (Fig. 2). The difference in sign reflects the reversal of the Bohr shift in *Limulus*; while the addition of acidic metabolites to the blood lowers the haemocyanin–oxygen affinity in a crustacean, the same phenomenon raises the haemocyanin–oxygen affinity in *Limulus* (Fig. 2).

The reversal of properties at a molecular level is accompanied by a similar reversal at the systemic level. When *Libinia* experiences hypoxic water ($P_{O_2} \leq 50$ Torr), blood pH goes up (Burnett, 1979). Although the mechanism responsible for the increase in pH in this species has not been investigated, in *Carcinus maenas* it appears to involve both hyperventilation accompanied by a slight tachycardia, and decreased carbon dioxide production during metabolism (deFur & Mangum, 1979; Burnett & Johansen, 1981). The result of the pH increase is an enhanced oxygen affinity of the blood, which is thus able to take up the same amount of oxygen at the gill in hypoxic water as in normoxic water, even though the partial pressure of oxygen in the blood is lower during hypoxia (Burnett, 1979).

When *Limulus* is exposed to hypoxic water, the frequency of the ventilatory movements of the opisthosomal appendages increases. Hyperventilation is counteracted in this species by a pronounced bradycardia and concomitant increases in blood lactate and the partial pressure of carbon dioxide; the resulting decrease in pH causes the oxygen affinity of the blood to go up (Towle *et al.*, 1982). The mechanisms are fundamentally different in the two groups of arthropods but, owing to the opposite Bohr shifts, the net results are the same – facilitation of oxygen uptake at the gill.

The only information available for the role of the excretory organ in acid–base regulation in a marine crustacean is very old and not very precise. It suggests that the antennal gland plays little or no role (Table 1) and, by implication, that the major site of acid output is the gill. The only recent information was obtained for a freshwater population of the crab *Callinectes sapidus* (Table 1), and also indicates that the antennal gland is not a site of acid–base regulation.

Clearly, the data are too few to justify confidence in a general statement about the role of the excretory organ in crustaceans. Equally clearly, the *Limulus* coxal gland is an important site of net acid output (Table 1). Under normoxic and isosmotic conditions the urine is considerably acid to the blood and the accumulation of acidic metabolites in the blood is very small.

Brief exposure to water at a partial pressure of 50 Torr, which is below the range of the partial pressure of oxygen over which oxygen uptake is regulated, results in an increase in the partial pressure of carbon dioxide in the blood, from 3.8 to 11.5 Torr, and *considerably greater* changes in lactate (Table 1). Prolonged exposure to air results in a somewhat smaller change in lactate, a slightly

Fig. 2. The effect of pH on haemocyanin–oxygen (HcO_2) affinity in the crustacean *Libinia emarginata* (dashed line, from Burnett, 1979) and the chelicerate *Limulus polyphemus* (solid line, data collected by C. O. Diefenbach); 25°C, 0.05 M tris-maleate buffered seawater (*Libinia*) or physiological saline (*Limulus*).

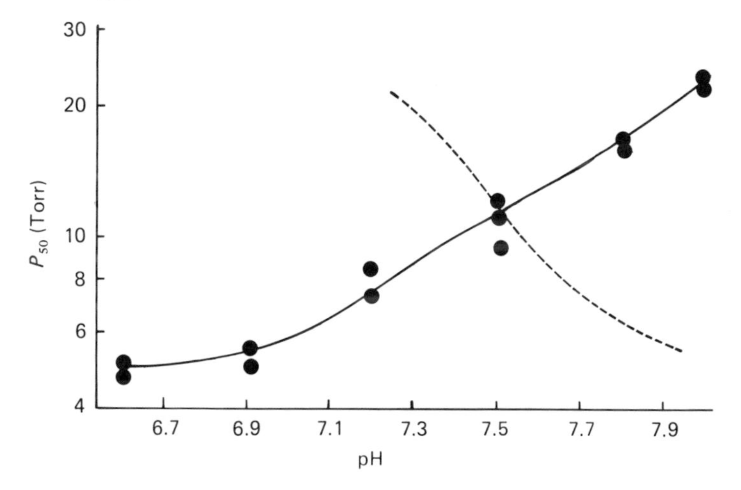

Table 1. *The relation between blood and renal acid–base status in aquatic arthropods*

	pH		HCO_3^- (mmol l^{-1})		Lactate (mmol l^{-1})	
	Blood	Urine	Blood	Urine	Blood	Source
Limulus polyphemus						
Water normoxic, 35‰	7.42	6.77	2.58	0.62	0.84	Towle *et al.* (1982)
Water normoxic, 8‰	7.47	7.77	3.91	4.06	?	Towle *et al.* (1982)
Water P_{O_2} 109 Torr, 2.5-h exposure, 35‰	7.40	?	2.09	?	1.40	Towle *et al.* (1982)
Water P_{O_2} 50 Torr, 2.5-h exposure, 35‰	7.31	?	7.03	?	2.0	Towle *et al.* (1982)
Air, 12–26-h exposure	6.91	6.61	1.77–2.26*	0.37	5.20	Towle *et al.* (1982) & *Johansen & Petersen (1975)
Homarus americanus						
Normoxic water, high salinity	7.45–7.60	7.40–7.55	?	?	?	Burger (1957)
Callinectes sapidus						
Normoxic and normocapnic freshwater	7.98	8.04	~5.3	5.3	?	Cameron & Batterton (1978)
Normoxic and hypercapnic (15 Torr) freshwater	7.65	7.97	12–13	9–13	?	Cameron (1978)

greater increase in the partial pressure of carbon dioxide, to as much as 14 Torr (Johansen & Petersen, 1975), and a large decrease in blood pH. The haemocyanin–oxygen affinity increases from about 11 to about 6 Torr. Although the urine does become more acid, the change is very small and therefore the blood–urine difference diminishes, suggesting that the ability of the coxal gland to acidify the product is approaching its limit. The greater change in pH during exposure to air, despite smaller changes in lactate, may be due to a decrease in net acid excretion. There is no change in the inorganic ion levels in the blood, which suggests that there is little output of water in any form. The large change in pH during air exposure may also be due to the accumulation of an unidentified acidic metabolite; a small increase (40–50 mOsmol kg^{-1}) in blood osmolality cannot be fully explained by the observed changes in carbon dioxide and lactate.

Anaerobic metabolism in *Limulus*, incidentally, appears to involve essentially the classical glycolytic pathway terminating in lactate (J. A. Fields, personal communication); no other end-products have been found in appreciable quantities. The only distinctive feature is the utilisation of D-lactate dehydrogenase (Long & Kaplan, 1973), presumably resulting in the D-form of the end-product.

The role of the gill in the tridacnid clams

The lamellibranch molluscs are often cited as examples of animals in which the primary function of the gill has shifted from gas exchange to filter-feeding. Water flows across these gills at high velocities and very little of the available oxygen in the current is taken up into the animal's tissues. The process of ventilation is not wasteful, however, because it brings only the requisite amounts of food to the gill and food replaces oxygen as the commodity in short supply (Hazelhoff, 1938; Jørgensen, 1952).

This relation is somewhat different, however, in at least one lamellibranch species, which contains high concentrations of haemoglobin in its blood (Mangum, 1980). One would expect it to differ as well in species such as the giant clams of the tropical Pacific seas that belong to the family Tridacnidae. Although these clams live in waters with little plankton available for filtration, they grow to enormous sizes, which is perhaps possible because they have abandoned filter-feeding in favour of a symbiotic relation with a dinoflagellate (Yonge, 1975). Specially adapted hyaline structures in the mantle of the clam focus sunlight onto large numbers of algal cells, which are positioned in the blood sinuses just within the mantle. For many years it has been recognised that the symbionts supply the host with soluble end-products of the photosynthesis of carbohydrates, and that the net oxygen balance during the day is positive; the clam and its green symbionts together produce more oxygen than they consume. At night, however, the oxygen balance becomes negative, as in a conventional filter-feeder (reviewed by Yonge, 1975).

While the nutritional importance of harbouring algal cells in animal tissues has been appreciated for some time, the respiratory consequences are only beginning to be explored. When exposed to hypoxic water in the light, for example, the symbionts found in the tissues of the sea anemone *Anthopleura* supply enough oxygen to eliminate an oxygen debt (Shick & Brown, 1977).

Since the food source in tridacnids is essentially endogenous, we wondered whether the respiratory profile continues to resemble that of filter-feeders. In the dark, *Tridacna squamosa* behaves differently from many lamellibranchs in that this species is an oxyconformer; oxygen uptake changes linearly with the ambient partial pressure of oxygen. The ventilation rate is only one-fifth of that found in the filter-feeder *Modiolus demissus*, and *T. squamosa* extracts an average of 42% of the available oxygen (Mangum & Johansen, 1982). The partial pressure of the oxygen in the blood taken from the heart is slightly higher than that in the blood in the adductor muscle sinus, which means that at least some of the oxygen consumed is carried in the blood from the gills and mantle to deep tissue; this is also true of *M. demissus* (Mangum, 1980).

Although the ventilation rate decreases in daylight, we wondered why it continues at all if both food and oxygen are produced endogenously. A number of possible answers come to mind, including a requirement for nutrients that cannot be supplied in adequate amounts by photosynthesis, such as the precursors of proteins. Perhaps less obvious but more important is the unique respiratory requirement of a source of carbon dioxide and a sink for oxygen.

In daylight, all of the relations that a respiratory physiologist expects to find in an aerobic animal are reversed in *T. squamosa*. The volume of oxygen in exhaled water is greater than the volume in water entering the siphons. The partial pressure of the oxygen in prebranchial blood exceeds that in postbranchial blood. Most importantly, the average partial pressure of the oxygen in the blood in samples taken from the adductor muscle was 162 Torr on a day when the barometric pressure was unexceptional, and the range included considerably higher values (Mangum & Johansen, 1982). The blood is slightly supersaturated with air, raising the possibility of oxygen toxicity* and the formation of embolisms in the tissues. The tridacnid gill, then, must function to carry out gas exchange in reverse.

The structure and function of the annelid gill

The annelids are among the simplest animals with an organ which zoologists have always called a gill. Annelid gills are, in fact, a bewildering array of structures whose functional organization is poorly understood. In the more primitive members of the phylum the gill is a simple, finger-like evagination of the

*The possibility of detoxification by the enzyme superoxide dismutase is a point of considerable interest.

body wall, without branches to increase its surface area or internal partitions to reinforce the separation of efferent and afferent flow (e.g. Fig. 4.6 in Mangum, 1976*a*). In this example (*Glycera*), coelomic fluid circulates through the gill and a closed circulatory system is absent. Gills of the same simple form, however, may be supplied by a loop of the closed circulatory system, as in the oligochaete *Alma nilotica* (Stephenson, 1930). In the more advanced families great elaborations of the external morphology of the gill have been described by taxonomists for more than a century; the degree to which the internal structure either is or is not highly organised has not been as widely appreciated.

Recently, the general features of both the more and the less highly organised gills were succinctly summarised by Kennedy (1979). A classification based on the extent to which the circulatory system is 'closed' within the gill, or the integrity of the epithelial lining of the vessels, has also been devised by Storch & Alberti (1978). The present discussion is focused on three examples: (1) the highly organised gill found in the lugworm *Arenicola cristata*, a species which is only moderately specialised for its burrowing, infaunal way of life; (2) the less highly organised gill of the sabellid polychaete *Eudistylia vancouveri*, a species which is more highly adapted for its tubicolous, epifaunal habitat; and (3) the parapodium of the relatively unspecialised species *Nereis succinea*. The nereid parapodium is not called a gill, and it has many functions other than gas exchange.

Arenicola

A. cristata has eleven pairs of branching filamentous gills which are arranged metamerically and positioned dorsally in the midregion of the body. The main trunk of a gill divides first into 6–11 primary branches (Fig. 3*a*) which are each about 1 mm long and 0.3 mm in diameter. Each primary branch subdivides to form secondary branches, then tertiary branches (about 0.14 mm in diameter) and finally quaternary or terminal branches (0.6 mm long and 0.07–0.10 mm in diameter). The surfaces of the terminal branches (Fig. 3) account for most of the gill surface area.

The afferent branchial vessel arises from the ventral vessel in the coelomic cavity and extends into the gill, branching in the same pattern as the gill rami, and then loops over at the tips of the terminal branches to return as the efferent branchial vessel. The afferent and efferent branchial vessels are connected by numerous transverse connecting or bridge vessels, which occur in the terminal branches about once every 35–40 μm. These bridges may also occur in the higher order branches, but they are never as frequent. Since the diffusion distance from the ambient medium to the branchial vessels is about 13–16 μm and the distance separating the bridges from the medium is 2–4 μm, it is clear that the bridges are the primary sites of gas exchange.

The gills of a number of polychaete families are organised in this way (Spies,

1973; Kennedy, 1979; Wells, Jarvis & Shumway, 1980), as are filamentous gills in other phyla such as the crustaceans (Balss, 1938). As pointed out by Storch & Alberti (1978) and also by Kennedy (personal communication), they differ in the important respect of whether or not the coelom extends into the filaments. In *A. cristata* it does not, but in other species the branchial vessels are also separated from the medium by a coelomic space of considerable size, which probably further diminishes gas exchange between them and the medium (see also below). The mechanical significance of this distinction has not been investigated, but it

Fig. 3. The gill of the annelid lugworm *Arenicola cristata*. (*a*) Scanning electron micrograph showing the trunk region (T), which is connected to the body, and several primary branches (PB). (*b*) Scanning electron micrograph showing primary (PB), secondary (SB), tertiary (TB) and terminal or quaternary (QB) branches. (*c*) Transmission electron micrograph of cross- and longitudinal sections through terminal branches showing afferent and efferent branchial vessels (BV) and bridges or connecting vessels (CV) between them. Magnification: ×108. (*d*) Higher magnification (×1552) longitudinal section through terminal branch, showing diffusion distance between connecting vessel (CV) and ambient medium (M). (Unpublished study by Jerome M. Johnson.)

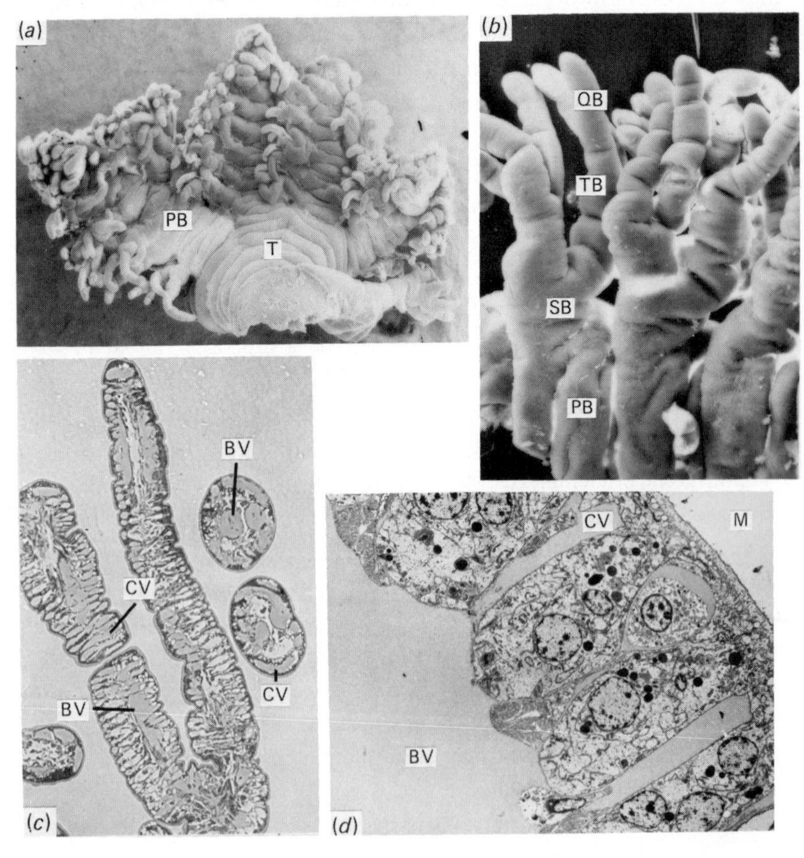

may prove to be related to the contractility of the gill and the presence of other skeletal elements. In many families the gill contracts rhythmically, possibly aided by movements of coelomic fluid, and thus functions as an auxiliary pump (Mangum *et al.*, 1975). In other families the erect posture of the gill is supported by endoskeletal cartilage (e.g. Person & Matthews, 1967) and the gill remains in a more or less fixed position so long as the animal is not disturbed.

The contractility of the gill may also be related to the size and frequency of infolding or the formation of macrovilli along the branchial epithelium, which appears to be more highly developed in the terebellids (Wells *et al.*, 1980) than in *Arenicola* (Fig. 3*b*).

The annelid cuticle is often penetrated by microvilli reaching through it to make contact with the medium (e.g. Kryvi, 1972; Mangum, Saintsing & Johnson, 1980). The cuticle itself is more or less smooth between the microvilli, so that the spaces between the microvilli do not contain unstirred fluid, as they do in a model transporting epithelium. Although the spaces between the branchial macrovilli are larger and less regular in shape than the microvillar spaces, it is not yet clear that their geometry changes enough during the contractions of the gill to cause significant fluid convection; thus their contribution to the gas exchange surface cannot be gauged.

The filamentous gill of the lugworms and many other species is organised inside both to effect the separation of oxygenated and deoxygenated blood, although the separation is not complete, and to assist in blood movements. However, the respiratory consequences of the rather elaborate internal organisation are mitigated by two features of the lugworms and their burrowing way of life. First, the gills are distributed metamerically along the length of the midregion. The diameter of the afferent and efferent branchial vessels is considerably smaller than the diameter of the enormous longitudinal vessels traversing the main axis of the body. Even in the posterior branchial segments where the afferent branchial vessel arises from the large ventral longitudinal vessel and the efferent branchial vessel empties into the large dorsal longitudinal vessel, most of the blood must bypass any particular pair of gills. In the more anterior branchial segments the afferents still arise from the ventral vessel, but the efferents now empty into the gut sinus, which must be a major target organ. So the dorsal vessel, one of the two major avenues for the blood supply to the tissues, must contain a large volume of blood that has not passed through the elaborate network in the gills. Not surprisingly, the partial pressure of oxygen in the blood flowing in both of the large longitudinal vessels is very low (Toulmond, 1973; Mangum, 1976*b*), and the oxygenation properties of the carrier molecule are highly adapted to function in such a partial pressure regime. Specifically, the haemoglobin–oxygen affinity is very high and the cooperativity of oxygen-binding very great. This means that a small change in the partial pressure of oxygen causes a large change in oxygenation, so that the blood does not unload very much at the partial pressure level in the dorsal vessel, but it retains very little of its load when returning from the tissues at the partial pressure in the ventral vessel, which is only a little lower (Mangum, 1976*b*).

The second feature that mandates a high-oxygen-affinity carrier despite the adaptations ensuring little blood mixing in the gill is the partial pressure of oxygen in the microenvironment of the burrow. It is appreciably lower than the partial pressure of oxygen in the air-saturated water outside. Lugworms usually pump water through the burrow in an anterior direction so that, by the time the current reaches the branchial segments in the midregion, the partial pressure of oxygen in the blood is somewhat lower than it would be if the blood equilibrated to normoxic water.

Eudistylia

On the other hand, it is somewhat surprising that the gill is not as well designed in another family, which is usually regarded as more advanced in other respects than the lugworms and more highly adapted to a tubicolous way of life. Members of the family Sabellidae inhabit vertical tubes, either within a soft sediment or, as in *Eudistylia vancouveri*, attached to a hard substratum. Structures such as the gills and the nephridia are no longer arranged metamerically, but instead are restricted to the anterior end, where materials can be most easily exchanged with the medium and flushed away from the vicinity of the tube. When undisturbed, the gill remains protruded into the water column from the anterior end of the tube in a more or less stationary position, showing no sign of frequent rhythmic contractions. The gill, a pinnate structure known as the branchial crown, has become a filter-feeding organ. It can, however, be withdrawn into the tube for protection.

In *E. vancouveri* the gill consists of two main trunks or branchioles, which immediately divide into numerous long, ciliated pinnae. These in turn branch into a double row of filaments called pinnules. A single branchial crown is composed of about 54 000 pinnules, the surfaces of which account for more than 70% of the surface area of the body, which is 30 cm^2/g wet wt (Vogel, 1980).

Each branchiole contains a single blood vessel (Fig. 4) that branches according to the plan described above, and ends blindly in the pinnules. Thus the internal organisation makes no provision for the separation of afferent and efferent blood flow, and the flow pattern is believed to be tidal (Fox, 1938; Ewer, 1941). The design is extremely simple and there is no sign of any adaptations to enhance gas exchange (Fig. 4). Moreover, the single blood vessel is enclosed within a coelomic space, so that the diffusion distance in an expanded gill is at least 10–12 μm (Fig. 4*b*).

Fig. 4. The gill of the annelid feather duster worm *Eudistylia vancouveri*. (*a*) Transmission electron micrograph of a cross-section of a contracted pinnule, showing the branchial vessel (BV) and the coelomic space (C). Magnification: ×2225. (*b*) Light micrograph showing expanded pinnule on left, and semi-contracted pinnule on right. Magnification: ×674.

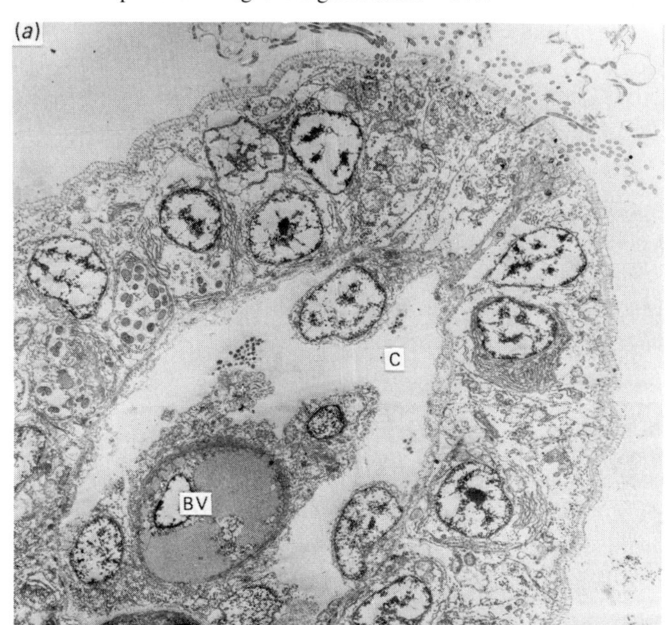

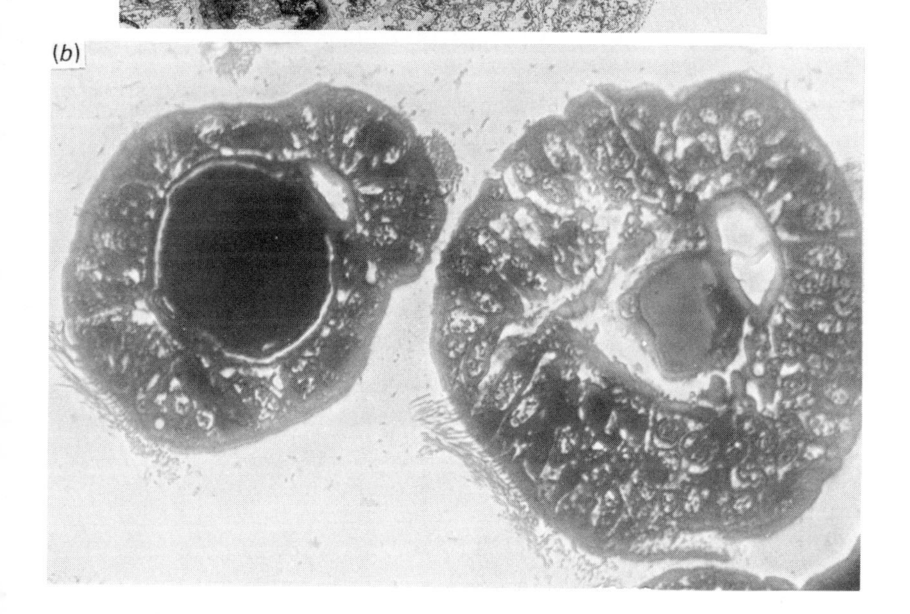

There is, however, reason to believe that the non-metameric design of the circulatory supply to the gills and the mode of feeding have resulted in a very different pattern of oxygen uptake in sabellids. First, the rate of ventilation can be up to three times higher in sabellids (Dales, 1957, 1961) than in arenicolids (Mangum, 1976*b*), and the percentage of oxygen extracted from the water current is only 6–10 in sabellids instead of 30–60 in arenicolids (Dales, 1957, 1961; Mangum, 1976). This indicates that as in other filter-feeders, it is food rather than oxygen which is the item in short supply in the water-current. The average partial pressure of the oxygen in the water with which the branchial blood in sabellids equilibrates, then, is quite high and, although it has not been measured, the partial pressure of postbranchial blood is believed to be high. In addition, observations on living animals made by Ewer (1941) suggest that the blood may pass into and out of the gill several times before reaching the ventral vessel, which is the main avenue of distribution to the tissues; the sabellid gill may be a rare example in the animal kingdom of a multiple pass exchange system.

Possibly for these reasons, the oxygen transport system seems to be one that operates at a high partial pressure of oxygen with large molecular gradients driving oxygen from the blood to the mitochondria. This hypothesis remains to be tested by the necessary measurements of the partial pressure of the oxygen in the blood, which have not been accomplished for technical reasons. The hypothesis is based on two lines of indirect evidence: (1) The oxygen affinity of the sabellid respiratory pigment, chlorocruorin, is invariably low, ranging from 44 Torr in *Myxicola* (Wells & Dales, 1975) to 52 Torr in *Spirographis* (Antonini, Rossi-Farelli & Caputo, 1962), at pH 7.4 and 15 °C. In *E. vancouveri*, P_{50} is 50 Torr

Fig. 5. The effect of exposure to carbon monoxide on oxygen uptake in *Eudistylia vancouveri*. 15°C, 31–33 ‰. Figures in brackets are the number of observations. (From Vogel, 1980.)

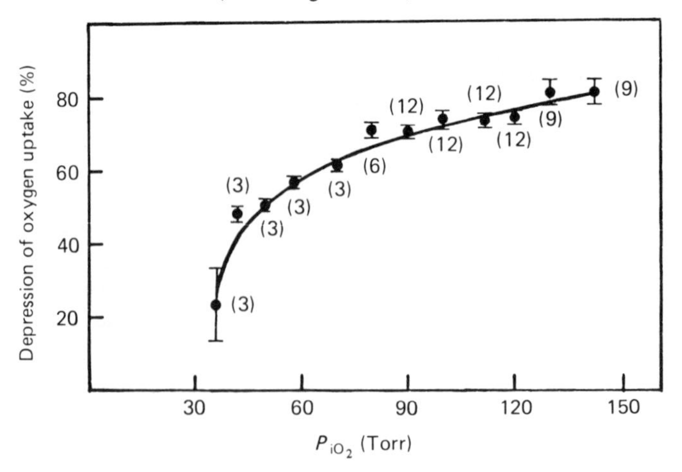

at 20 °C (R. C. Terwilliger, peronal communication), which would be about 44 Torr at 15 °C and pH 7.4, the measured value in postbranchial blood (Vogel, 1980). And yet, when the oxygen binding site is blocked with carbon monoxide, total oxygen uptake decreases by 60–80% but only at a high ambient partial pressure of oxygen (Fig. 5). The only way that a low-oxygen-affinity carrier can transport 60–80% of the oxygen consumed is to become highly oxygenated at the gill, which can occur only at a high partial pressure of the oxygen in the blood, and to remain partially oxygenated until the blood reaches most of the delivery sites at the tissues, which can also occur only at a relatively high pre-branchial partial pressure of the oxygen in the blood.

The overall function of the annelid respiratory system may be determined, then, by factors other than the microcirculation in the gill.

Nereis succinea

Members of the genus *Nereis* have no structures called gills. On the basis of very astute observations on living animals, however, Paul Nicoll, the distinguished investigator of the mammalian microcirculation, recognised in 1954 that the parapodia function as sites of oxygen uptake into the blood.

Nicoll described two types of capillaries. One type ends blindly at the termini of blood vessels that branch into the several ventral ligules of the parapodium; he referred to beds of these as the systemic circulation. The other type 'clearly satisfies the classical concept as being simple endothelial tubes that are the terminal anastomotic arborisations of larger vessels' (1954, p. 75). He called these beds of flow-through capillaries, which are confined to the dorsal ligule, known as notopodial ligule 1, and to the adjacent body wall, the respiratory circulation. Only in *N. succinea* (= *limbata*) are the two circulatory routes completely separated; in *N. virens* they are connected to the large longitudinal vessels by a single trunk.

More recent studies of the capillary circulation in nereids, and also in other families, have been concerned in large part with features that are not critical to its role in respiratory gas exchange, such as the nature of the cellular lining. Most investigators disagree with Nicoll's (1954) conclusion that the lining is correctly designated an endothelium, although the argument is in part semantic (Kryvi, 1972; Storch & Welsch, 1972; Nakao, 1974).

Our studies of the capillary microcirculation in several families support Nicoll's (1954) view of its function, and they also emphasise several aspects that have not been widely recognised. First, the systemic circulation is sparse. Despite a thorough search, we have been unable to find, in any family examined, a highly organised systemic circulation to deep tissue such as the two cylinders of opposing muscles which make up the digestive tube and the body wall. In *Arenicola cristata*, they are simply absent. Capillaries lie on the coelomic face of the digestive tube of younger animals belonging to several families (Spies, 1973; Nakao, 1974), but they fuse to form a continuous sinus as the animal ages. Perhaps the closest approximation to deep capillary beds is the blood supply to the oesophageal glands in lugworms, which in fact are not very thick tissues, and the capillaries are actually unlined tubular sinuses (J. M. Johnson, unpublished observations).

In nereids the systemic capillary circulation is essentially confined to the parapodia. Ewer (1941) mentions the presence of capillaries between the cells of the gonads of serpulids, but elsewhere the capillaries ramify on the surfaces of organs. Capillaries, in the form of loop vessels, are present only in the outer (circular) muscle layer in the gill-less maldanids (Pilgrim, 1966), where they appear to lie within the limiting diffusion distance from the medium and thus may prove to be respiratory rather than systemic.

The oxygen supply to tissue that lies further than the limiting diffusion distance from the medium must be carried largely by coelomic fluid which, in most species, has a very low oxygen-carrying capacity. For example, at the microenvironmental partial pressure of oxygen, the partial pressure of oxygen in the coelomic fluid of *A. cristata* is less than 50 Torr and free oxygen is not supplemented by carrier-bound oxygen. Either the metabolic rate is very low in those tissues, which is not true when they are excised, or it is severely limited by the oxygen supply.

Secondly, our observations with both the scanning and transmission electron microscopes strongly support the view that the role of the superficial capillaries in the nereid parapodia is gas exchange with the medium.

The structure of the nereid parapodia changes gradually from the anterior to the posterior end. Anteriorly, the parapodia are divided into the dorsal notopodium, which has three ligules, and the ventral neuropodium, which has two. In both regions the respiratory microcirculation is essentially confined to the uppermost notopodial ligule and to a patch of dorsal body wall immediately adjacent to it (Fig. 6*c*). Posteriorly, notopodial ligule 1 enlarges greatly to a flattened strap-like structure and the other two notopodial ligules either disappear or become greatly reduced (Fig. 6). Clearly, these are the more important respiratory parapodia.

Blood enters the capillary bed in the ventrolateral notopodial vessel (Fig. 6*b*), passes through the bed, exits from the dorosolateral vessel (Fig. 6*c*), and continues to a second bed of flow-through capillaries on the dorsal surface of the body wall (Nicoll, 1954). The notopodial capillaries lie immediately beneath the cuticle, so that the diffusion distance from the medium to the blood ranges from only 1.5 to 2.3 μm (Fig. 6*d*).

Even though its respiratory role is not so obvious that it was designated a gill

Fig. 6. The parapodium of the annelid sandworm *Nereis succinea*. (*a*) Scanning electron micrograph of the anterior parapodium, showing the notopodial (NoLi) and neuropodial (NeLi) ligules. (*b*) Macrophotograph of a posterior parapodium, showing the afferent or ventrolateral (VLV) vessel supplying the bed of flow-through capillaries or connecting vessels (CV) in notopodial ligule 1. Magnification: ×16–20. (*c*) Macrophotograph of a posterior parapodium showing the efferent or dorsolateral (DLV) vessel which collects the blood from capillaries (CV) in notopodial ligule 1. Magnification: ×16–20. (*d*) Scanning electron micrograph of the posterior parapodium, as in (*c*), showing restriction of the superficial capillaries to notopodial ligule 1 (NoLi 1). Magnification: ×35. (*e*) Transmission electron micrograph showing a cross-section through the respiratory microcirculation. Flow-through capillaries (CV) are closely apposed to the cuticle (Cu). Magnification: ×1877.

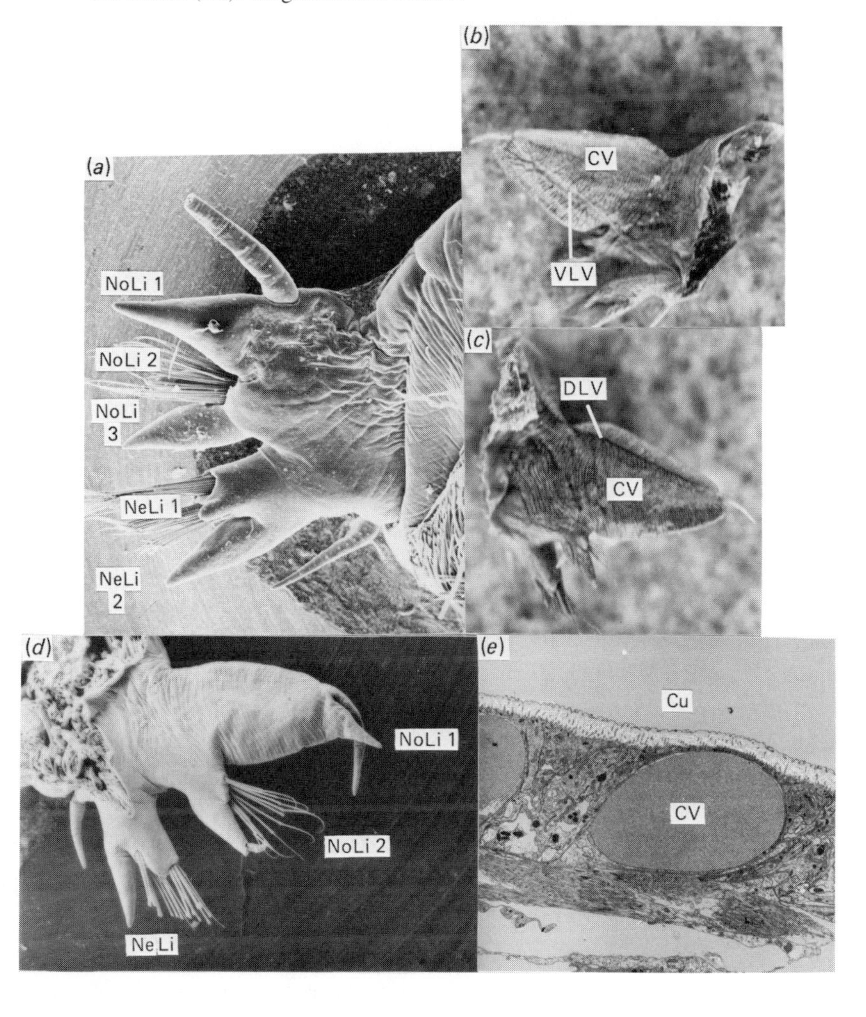

by the early morphologists, the upper notopodial ligule appears to have at least one of the other functions that we associate with gills. Below about 8–12$^{0}/_{00}$, *N. succinea* actively absorbs sodium chloride from the medium in order to maintain a hyperosmotic condition of its body fluids. When the body is slit longitudinally, the digestive tube removed and the body wall mounted as a rather thick sheet in an Ussing-type chamber, the electrical potential across the preparation shifts from insignificant at 18$^{0}/_{00}$ to negative at 3.5–7$^{0}/_{00}$, and the preparation transports labelled sodium ions inwards (Doneen & Clark, 1974). The shift from osmoconformity to osmoregulation in an intact animal is accompanied by increases in oxygen uptake and ammonia excretion, both of which are sensitive to the cardiac glycoside ouabain, a specific inhibitor of the sodium- and potassium-activated ATPases. The evidence suggests the participation of these enzymes in the uptake of sodium ions and the output of ammonium ions (Oglesby, 1975; Mangum, Dykens, Henry & Polites, 1978).

At 35$^{0}/_{00}$ no activity of the enzymes can be detected in strips taken from both the dorsal and the ventral surfaces of the body wall and cleaned of nephridial and, as much as possible, muscle tissue (Mangum *et al.*, 1980). With the single exception of the upper notopodial ligule, in which the activity is very small, none can be detected in homogenates of the various parapodial ligules taken from both the anterior and posterior regions.

We cannot dissect the capillary bed free of abundant amounts of parapodial tissue such as the motor muscles and setae, and ATPase activity is expressed per unit of protein. With so much irrelevant protein, we would expect the value for ATPase activity to be lower than in a real gill, which it is (Mangum *et al.*, 1980). But at 5–8$^{0}/_{00}$ we would also expect to be able to detect an increase in activity at the site of uptake of sodium ions and output of ammonium ions. No activity could be detected in body-wall strips, even when they were not cleaned of muscle and included the bed of flow-through capillaries immediately adjacent to the parapodia (Mangum *et al.*, 1980). Significant activity, however, was found in the parapodia. When we dissected the parapodia into their various ligules, we found that all of the activity resided in the upper notopodial ligule. Most interestingly, the anterior notopodia, which are less important in gas exchange, had more activity than the posterior notopodia. It was, in fact, the anterior portion of the body that had been used to demonstrate the non-equilibrium electrical potential and the uptake of labelled sodium ions (Doneen & Clark, 1974). Thus the anatomical separation of respiratory and osmotic exchange surfaces appears to have been begun already in the annelids.

The following individuals made substantial contributions to the investigations described above: D. G. Saintsing and D. W. Towle, University of Richmond;

K. Johansen, University of Aarhus; C. O. Diefenbach, J. M. Johnson, N. A. Mauro, J. Thomas and G. M. Vogel, College of William and Mary. The author was supported by grants from the US National Science Foundation, Regulatory Biology Program (PCM 77–20159 and 80–18709).

References

Antonini, E., Rossi-Farelli, A. & Caputo, A. (1962). Studies on chlorocruorin. I. The oxygen equilibrium of *Spirographis* chlorocruorin. *Archives of Biochemistry and Biophysics,* **97,** 336–42.

Balss, H. T. (1938). Stomatopods. In *Klassen und Ordnungen des Tierreichs,* vol. 5, Section 1, bk 6, pt II, ed. H. G. Bronns, pp. 36–116. Leipzig.

Burger, J. W. (1957). The general form of excretion in the lobster, *Homarus. Biological Bulletin of the Marine Biological Laboratory, Woods Hole,* **113,** 207–23.

Burnett, L. E. (1979). The respiratory function of hemocyanin in the spider crab, *Libinia emarginata,* and the ghost crab, *Ocypode quadrata,* in normoxia and hypoxia. *Journal of Experimental Zoology,* **210,** 289–300.

Burnett, L. E. & Johansen, K. (1981). The role of branchial ventilation in blood acid–base changes in the shore crab, *Carcinus maenas,* during hypoxia. *Journal of Comparative Physiology,* **141,** 489–94.

Cameron, J. N. (1978). Effects of hypercapnia on blood acid–base status, NaCl fluxes, and trans-gill potential in freshwater blue crabs, *Callinectes sapidus. Journal of Comparative Physiology,* **123,** 137–41.

Cameron, J. N. & Batterton, C. V. (1978). Antennal gland function in the freshwater blue crab, *Callinectes sapidus*: water, electrolyte, acid–base balance and excretion. *Journal of Comparative Physiology,* **123,** 143–8.

Copeland, D. E. & Fitzjarrell, A. T. (1968). The salt absorbing cells in the gills of the blue crab (*Callinectes sapidus* Rathbun) with notes on modified mitochondria. *Zeitschrift für Zellforschung und Mikroskopische Anatomie,* **92,** 1–22.

Dales, R. P. (1957). Some quantitative aspects of feeding in sabellid and serpulid worms. *Journal of the Marine Biology Association, UK,* **36,** 309–16.

Dales, R. P. (1961). Observations on the respiration of the sabellid polychaete *Schizobranchia insignis. Biological Bulletin of the Marine Biological Laboratory, Woods Hole,* **121,** 82–91.

de Fur, P. L. & Mangum, C. P. (1979). The effects of environmental variables on heart rates of invertebrates. *Comparative Biochemistry and Physiology,* **62A,** 283–94.

Doneen, B. A. & Clark, M. E. (1974). Sodium fluxes in isolated body walls of the estuarine polychaete, *Nereis* (*Neanthes*) *succinea. Comparative Biochemistry and Physiology,* **48A,** 221–8.

Evans, D. H. (1975). Ion exchange mechanisms in fish gills. *Comparative Biochemistry and Physiology,* **51A,** 491–6.

Ewer, D. W. (1941). The blood systems of *Sabella* and *Spirographis. Quarterly Journal of the Microscopical Society,* **82,** 587–620.

Fox, H. M. (1938). On the blood circulation and metabolism of sabellids. *Proceedings of the Royal Society of London, Series B,* **112,** 479–95.

Hazelhoff, E. H. (1938). Uber die Ausnutzung des Sauerstoffs bei verschiedenen Wassertieren. *Zeitschrift für Vergleichende Physiologie,* **26,** 307–27.

Johansen, K. & Petersen, J. A. (1975). Respiratory adaptations in *Limulus*

polyphemus (L.). In *Physiological Ecology of Estuarine Invertebrates,* ed. F. J. Vernberg, pp. 129–46. Columbia: University of South Carolina Press.
Jørgensen, C. B. (1952). On the relationship between water transport and food requirements in some marine filter feeding invertebrates. *Biological Bulletin of the Marine Biological Laboratory, Woods Hole,* **103,** 356–63.
Kennedy, B. (1979). Blood circulation in polychaete gills. *American Zoologist,* **19,** 868.
Krogh, A. (1939). *Osmotic Regulation in Aquatic Animals.* New York: Dover Publications.
Krogh, A. (1941). *The Comparative Physiology of Respiratory Mechanisms.* New York: Dover Publications.
Kryvi, H. (1972). Fine structure of the blood capillaries in the ventral mucous gland of *Sabella penicillus* L. (Polychaeta). *Sarsia,* **49,** 59–64.
Lochhead, J. N. (1950). *Xiphosura polyphemus.* In *Selected Invertebrate Types,* ed. F. A. Brown, pp. 360–82. New York: John Wiley and Sons.
Long, G. O. & Kaplan, N. O. (1973). Diphosphopyridine nucleotide linked D-lactate dehydrogenases from the horseshoe crab *Limulus polyphemus* and the seaworm *Nereis virens. Archives of Biochemistry and Biophysics,* **154,** 696–725.
Mangum, C. P. (1976*a*). Primitive respiratory adaptations. In *Adaptations to the Environment: Essays on the Physiology of Marine Animals,* ed. R. C. Newell, pp. 191–278. London: Butterworth.
Mangum, C. P. (1976*b*). The oxygenation of hemoglobin in lugworms. *Physiological Zoology,* **49,** 85–99.
Mangum, C. P. (1980). Distribution of the respiratory pigments and the role of anaerobic metabolism in the lamellibranch molluscs. In *Animals and Environmental Fitness,* ed. R. Gilles, pp. 171–84. Oxford: Pergamon Press.
Mangum, C. P., Dykens, J. A., Henry, R. P. & Polites, G. (1978). The excretion of NH_4^+ and its ouabain sensitivity in aquatic annelids and molluscs. *Journal of Experimental Zoology,* **203,** 151–7.
Mangum, C. P. & Johansen, K. (1982). The influence of symbiotic dinoflagellates on respiratory processes in the giant clam *Tridacna squamosa. Pacific Science,* in press.
Mangum, C. P., Saintsing, D. G. & Johnson, J. M. (1980). The site of ion transport in an estuarine annelid. *Marine Biology Letters,* **1,** 197–204.
Mangum, C. P., Woodin, B. L., Bonaventura, C., Sullivan, B. & Bonaventura, J. (1975). The role of coelomic and vascular hemoglobins in the annelid family Terebellidae. *Comparative Biochemistry and Physiology,* **51A,** 281–94.
Nakao, T. (1974). An electron microscopic study of the circulatory system in *Nereis japonica. Journal of Morphology,* **144,** 217–36.
Nicoll, P. A. (1954). The anatomy and behaviour of the vascular systems in *Nereis virens* and *Nereis limbata. Biological Bulletin of the Marine Biological Laboratory, Woods Hole,* **106,** 69–82.
Oglesby, L. C. (1975). The effects of salinity changes on the metabolism of estuarine nereid polychaetes. *American Zoologist,* **15,** 795.
Person, P. & Matthews, M. B. (1967). Endoskeletal cartilage in a marine polychaete, *Eudistylia polymorpha. Biological Bulletin of the Marine Biological Laboratory, Woods Hole,* **132,** 244–52.
Pilgrim, M. (1966). The anatomy and histology of the blood system of the maldanid polychaetes *Clymenella torquata* and *Euclymene oerstedi. Journal of Zoology,* **149,** 261.
Sassaman, C. & Mangum, C. P. (1973). Adaptations to environmental oxygen

levels in infaunal and epifaunal sea anemones. *Biological Bulletin of the Marine Biological Laboratory, Woods Hole,* **143,** 657–78.
Shick, J. M. & Brown, W. I. (1977). Zooxanthellae-produced O_2 promotes sea anemone expansion and eliminates O_2 debt under environmental hypoxia. *Journal of Experimental Zoology,* **201,** 149–55.
Shuster, C. N. (1978). *The Circulatory System and Blood of the Horseshoe Crab* Limulus polyphemus *L.* 63 pp. Washington, D.C.: US Department of Energy.
Spies, R. B. (1973). The blood system of the flabelligerid polychaete *Flabelliderma commensalis* (Moore). *Journal of Morphology,* **139,** 465–90.
Stephenson, J. (1930). *The Oligochaeta.* 978 pp. Oxford: Clarendon Press.
Storch, V. & Alberti, G. (1978). Ultrastructural observations on the gills of polychaetes. *Helgolander Wissenschaftliche Meeresuntersuchungen,* **31,** 169–79.
Storch, V. & Welsch, V. (1972). Ultrastructure and histochemistry of the integument of air-breathing polychaetes from mangrove swamps of Sumatra. *Marine Biology,* **17,** 137–44.
Toulmond, A. (1973). Tide-related changes of blood respiratory variables in the lugworm *Arenicola marina* (L.). *Respiration Physiology,* **19,** 130–44.
Towle, D. W., Mangum, C. P., Johnson, B. A. & Mauro, N. A. (1982). The role of the coxal gland in ionic, osmotic and pH regulation in the horseshoe crab *Limulus polyphemus.* In *Physiology and Biochemistry of the Horseshoe Crab* Limulus polyphemus, ed. J. Bonaventura. New York: Alan Liss & Co. (In press.)
Vogel, G. M. (1980). Oxygen uptake and transport in the sabellid polychaete *Eudistylia vancouveri* (Kinburg). 32 pp. M.A. Thesis. College of William and Mary, Williamsburg, Virginia.
Wells, R. M. G. & Dales, R. P. (1975). The respiratory significance of chlorocruorin. In *Proceedings of the 9th European Marine Biology Symposium,* ed. H. Barnes, pp. 673–81. Aberdeen University Press.
Wells, R. M. G., Jarvis, P. J. & Shumway, S. E. (1980). Oxygen uptake, the circulatory system, and haemoglobin function in the intertidal polychaete *Terebella haplochaeta* (Ehlers). *Journal of Experimental Marine Biology and Ecology,* **46,** 255–77.
Yonge, C. M. (1975). Giant clams. *Scientific American,* **232**(4), 96–105.

KJELL JOHANSEN

Respiratory gas exchange of vertebrate gills

Vertebrate gills associated with the primary branchial arches can be internal, as in typical fish gills, or project externally, as in larval lungfishes and many amphibians. In neotenic amphibians external gills occur throughout the life cycle, for example among urodeles. Some vertebrates show gill-like filamentous structures which are not derived from the branchial arches, yet have alleged or demonstrated functions in respiratory gas exchange.

Gills are multipurpose organs which serve in the passive exchange of gases, ions and nitrogenous metabolic products. Gills can also be the site of active exchange or transport of agents such as ions.

The role and efficiency of gills in respiratory gas exchange will be reviewed in the following sections. The inherent oxygen requirement of gill tissues will also be discussed. The emphasis will be on oxygen uptake in fish gills for which more information is available than for gills from other vertebrates.

The body surface may substitute or assist gas exchange in gills

Embryos and larvae

Before gills develop in fish larvae the body surface constitutes the only surface for external gas exchange. Fish larvae show many specialisations which assist cutaneous gas exchange by promoting convective movement of the surrounding water either passively by positioning the larvae in currents, or actively by using muscular fin movements (Liem, 1981), or, as described for lungfishes, by using a transient epidermic ciliary apparatus (Budgett, 1901; Whiting & Bone, 1980). Most interestingly in the lungfish (*Protopterus*), the external gills of newly hatched larvae have cilia on their external surface (Greenwood, 1958) which are an obvious aid to convection, and thus serve to increase gas diffusion across the gills prior to the development of a muscular apparatus for active movement of the gills or the water irrigating them. Also, in amphibians, the surface epithelium of embryos and larvae may be ciliated (Billett & Courtenay, 1973). A most interesting and perhaps specialised case was recently described for the larvae of an air-breathing teleost fish, *Monopterus* (Liem, 1981). Prior to gill development, large muscular and vascular fins, in addition to functioning as external gills, set up a water current directed posteriorly along the fish. An extensive capillary network below the epithelial surfaces on the fins and the yolk sac channels blood, producing a counter-current to the water flow, and making the larva analogous to a fish gill lamella. Liem offers experimental evidence for increased efficiency of oxygen removal from water when the counter-current exchange is in operation. He predicts that the mechanism allows *Monopterus* larvae to develop in hypoxic water and explore oxygen-richer surface water.

Fish with developed gills

The important contribution from skin gas exchange to the total need was first demonstrated by Krogh (1904) for eels, a finding subsequently extended to air-breathing fish (Lomholt & Johansen, 1979) and the haemoglobin-free ice-fish, *Chaenocephalus aceratus* (Hemmingsen & Douglas, 1970). More recent studies have revealed also that in fish of more ordinary habitats and behaviour, the skin may play a very significant role in gas exchange, contributing up to about 30% of the total for both freshwater and marine species in well-aerated water (Kirsch & Nonnotte, 1977; Nonnotte & Kirsch, 1978). Steffensen, Lomholt & Johansen (1981) have demonstrated that this important skin respiration also occurs in fish such as plaice (*Pleuronectes platessa*) which are normally buried in soft substrate. Most surprisingly, the importance of skin oxygen uptake relative to that of gills was increased in hypoxic water and made up 37% of total uptake at a water P_{O_2} of 40 Torr for *Pleuronectes* (Fig. 1).

Fig. 1. Total, cutaneous and gill oxygen uptake at different ambient oxygen tensions in plaice, *Pleuronectes platessa*. Mean ± 1 s.d. (From Steffensen *et al.*, 1981.)

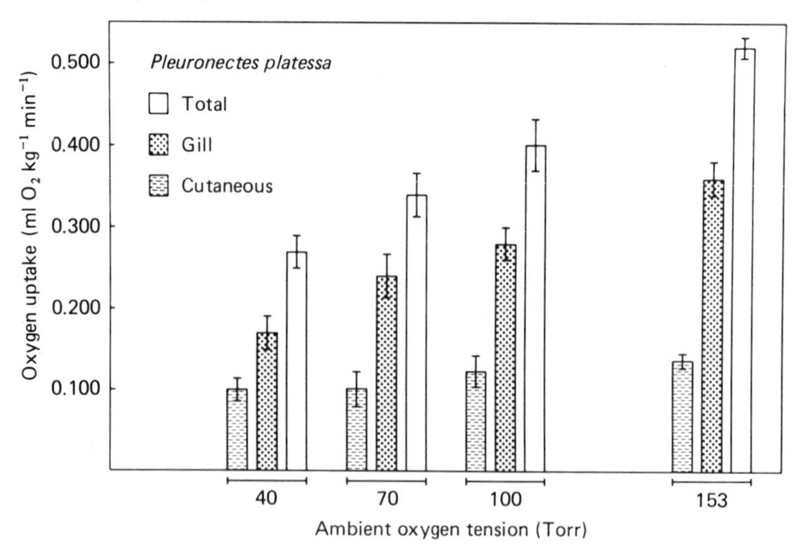

The performance of fish gills

The effectiveness of fish gills in gas exchange depends on the lamellar surface area, the water to blood diffusion distance and the magnitudes and matching of gill ventilation and blood flow. The efficiency can be quantitatively expressed by the transfer factor, T_{O_2}, or gill diffusing capacity, D_{O_2}, but the necessary measurements are difficult to make in intact fish.

If the exhaled water current can be confined, samples reflecting truly mixed exhaled water can be obtained, and gill ventilation can be measured continuously by, for instance, an electromagnetic flow-measuring technique. This method does not impose much restraint on the fish and the gill gas exchange performance can be assessed.

Fig. 2 shows this methodical approach applied on a carp, *Cyprinus carpio,*

Fig. 2. Carp, *Cyprinus carpio,* equipped with rubber mask permitting channelling of the respiratory water current for direct and continuous measurement of ventilation by an electromagnetic flow probe and for continuous monitoring of exhaled water P_{O_2}. (From Lomholt & Johansen, 1979.)

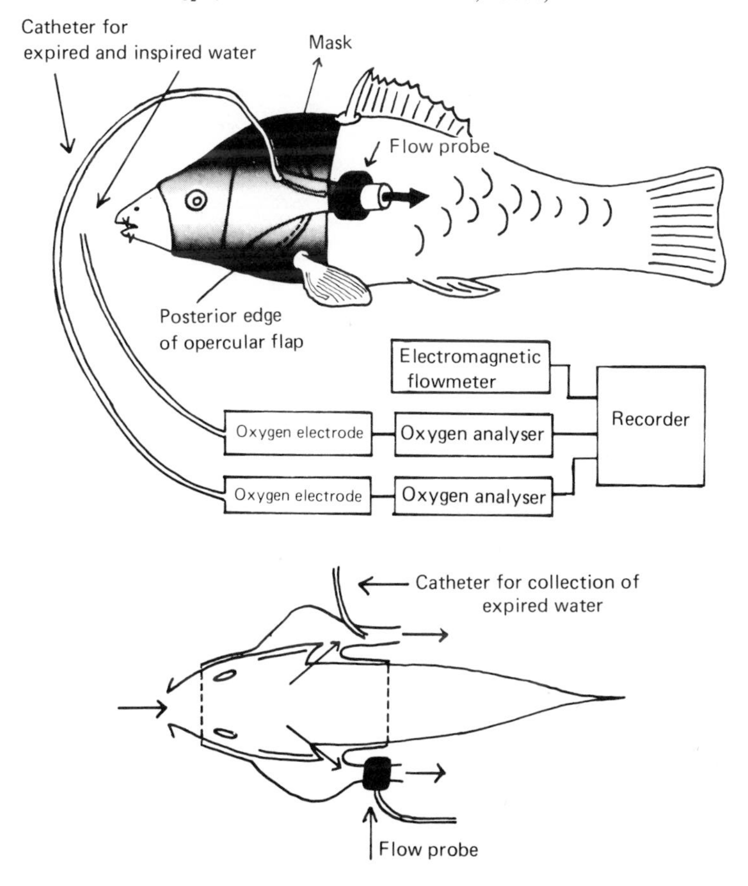

and Fig. 3 demonstrates how the ventilatory pattern changes from a periodic to a steady rhythm when the ambient water becomes hypoxic. The extraction of oxygen from the ventilatory current in carps is very high, exceeding 80%. This figure is most likely to be among the highest on record for fish (Fig. 3, Table 1).

Some fish, notably the fast-swimming pelagic species like the tunas, are able to maintain high oxygen extractions even at very high rates of gill ventilation (exceeding 3 l kg^{-1} min^{-1}). Stevens (1972) claims that tuna fish can extract 50% or more oxygen from the ventilatory current at ventilation rates many-fold those of other fishes. Not only the large surface areas of tuna gills, but also fusion of adjacent secondary lamellae at their tips make a sieve-like structure which eliminates dead space that would arise from separation of the filament tips at high rates of swimming, when gill gas exchange efficiency is maximally taxed in pelagic species (Muir & Kendall, 1968). Free-swimming skipjack tunas (*Katsowonus pelamis*) had oxygen extractions between 43% and 75%, with a mean value of 56%, during sustained swimming. Spot checks on arterial blood oxygen tensions showed values averaging 90 Torr, which clearly testifies to the operation of an efficient counter-current exchange during swimming.

Table 1 also gives values for the ventilatory requirement which expresses the need for gill water flow relative to oxygen uptake. This expression is also a useful measure of efficiency in gill oxygen exchange. The table shows the salmonid rainbow trout to have the lowest ventilatory requirement of the species

Fig. 3. Tracings of gill ventilatory flow in normoxia-acclimated carp at three different levels of ambient water P_{O_2}. (From Lomholt & Johansen, 1979.)

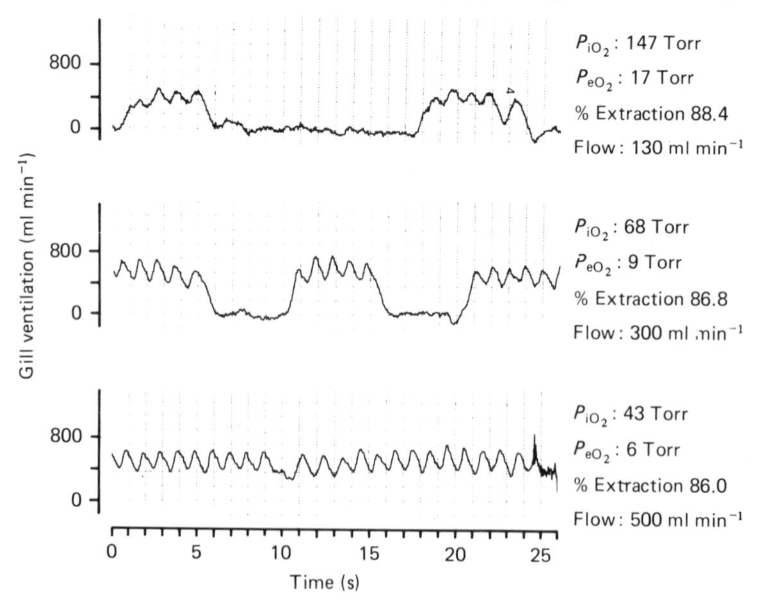

Table 1. *Comparison of ventilation, oxygen extraction and ventilatory requirement of different fish species*

Species	Weight (kg)	Temperature (°C)	Water oxygen tension, P_{iO_2} (Torr)	Ventilation, $\dot{V}_g$ (ml kg^{-1} min^{-1})	Oxygen extraction (%)	Oxygen consumption, $\dot{V}_{O_2}$ (ml kg^{-1} min^{-1})	Ventilatory requirement ($\dot{V}_g/\dot{V}_{O_2}$)	Source
Cyprinus carpio								
Carp (normoxia-acclimated)	~1	—	>100	195	80	0.80	244	Lomholt & Johansen (1979)
			< 40	1 122	72	1.03	1 089	
Carp (hypoxia-acclimated)	~1	—	>100	212	80	0.82	259	Lomholt & Johansen (1979)
			< 40	799	85	0.85	940	
Carp	0.174	20	high	327	72	1.17	280	Saunders (1962)
			low	3 065	46	1.17	2 620	
Carp	2–6	10	117–145	—	72–87	0.47	—	Garey (1967)
			67–103	—	52–80	0.47	—	
Ictalurus punctatus								
Catfish	0.004–0.009	24	~150	925	47	2.4	403	Gerald & Cech (1970)
Salmo gairdneri								
Rainbow trout	0.5–1	15	150	274	55	1.40	196	Holeton & Randall (1967)
			30	3.560	20	1.40	2 543	
Catostomus commersoni								
Sucker	0.250	20	high	336	56	0.99	339	Saunders (1962)
			low	12.960	7	0.99	13 090	
Ictaloros nebulosus								
Bullhead	0.186	20	high	290	68	0.99	293	Saunders (1962)
			low	7.419	13	0.99	7 494	

Table 1. (*continued*)

Species	Weight (kg)	Temperature (°C)	Water oxygen tension, P_{iO_2} (Torr)	Ventilation, $\dot{V}_g$ (ml kg^{-1} min^{-1})	Oxygen extraction (%)	Oxygen consumption, $\dot{V}_{O_2}$ (ml kg^{-1} min^{-1})	Ventilatory requirement ($\dot{V}_g/\dot{V}_{O_2}$)	Source
Squalus suckleyi								
Pacific dogfish	2–4.3	6–10	136	102 (51–170)	58 (42–76)	0.31	329	Hanson & Johansen (1970)
			123	299 (210–500)	37 (25–49)	0.51	586	
Scyliorhinus stellaris								
Dogfish	2.4–3.9	—	155	248	47	0.67	370	Piiper & Schumann (1967)
Dogfish	2–5	18	156 (rest)	320	64	1.10	291	Piiper *et al*. (1977)
			156 (swimming)	920	38	1.93	477	
Platichthys stellatus								
Starry flounder	1–8	9–11	~130	141.5	55	0.45	314	Watters & Smith (1973)
			~ 60	322.3	45	0.35	920	
Starry flounder	0.300–1.2	10	150	109.2	68	0.46	237	Wood *et al*. (1979)
Carassius carassius								
Goldfish	0.100	25	~150	450	48.3	1.25	360	Dejours *et al*. (1968)
Callionymus lyra								
Dragonet	0.100	11.5	~150	440	30.7	0.84	524	Hughes & Umezawa (1968)

Platichthys flesus								
Flounder (normoxia-acclimated)	0.150–0.500	8–10	150	88.6	76	0.45	197 (rest)	Kerstens *et al*. (1979)
Flounder (hypoxia-acclimated)	0.150–0.500	8–10	30	308	56	0.24	1 283	Kerstens *et al*. (1979)
Chaenocephalus aceratus								
Icefish	0.500–1.823	~0	150	197	18	0.54	372 (rest)	Holeton (1970)
Katsowonus pelamis								
Tuna (swimming)	1	23–25	150	~3 000	56	8.0	375	Stevens (1972)
Acipenser transmontanus								
Sturgeon	0.85–1.06	15	150	350	35–40	0.90–1.0	~350	Burggren & Randall (1978)
Neoceratodus fosteri								
Australian lungfish	4.2–8.1	18	315	315	50	0.50	630	Johansen *et al*. (1967)

tabulated. The carp similarly has a low ventilatory requirement, a factor giving a higher scope for ventilation increase during the frequently encountered hypoxic water.

Predictably, the fast-swimming fishes among both teleosts and elasmobranchs receive an important complement of their gill ventilatory requirement from ram ventilation. How much this may be and which fishes employ it is an obvious challenge for future studies to resolve. An important corollary of ram ventilation is its secondary effect of altering drag along the post-opercular body profile of the swimming fish (Freadman, 1981).

Gill oxygen uptake in hypoxic water

Most significantly, some species, like the carp, are able to maintain a high oxygen extraction even when the ventilation is increased five times or more during hypoxia. This implies an effective compensation in oxygen uptake during hypoxia and must involve a concurrent increase in the gill transfer factor (diffusing capacity) for oxygen brought about either by an increase in the effective lamellar surface area, or an alteration of blood flow and its distribution within the secondary lamellae. The physiological consequences of these changes must be a reduced P_{O_2} gradient across the gas exchange surfaces. Alterations in blood respiratory properties will also influence this gradient, either acutely, as a result

Fig. 4. Percentage extraction of oxygen from the ventilatory water current in normoxia- (filled circles) and hypoxia- (open circles) acclimated carp as a function of inspired oxygen tension. Each point is the average of 30–60 determinations in five different fish. Bars are 1 s.d. Except at the highest level of oxygen tension, average values for the two acclimation groups are significantly different (t-test, $P < 0.001$). (From Lomholt & Johansen, 1979.)

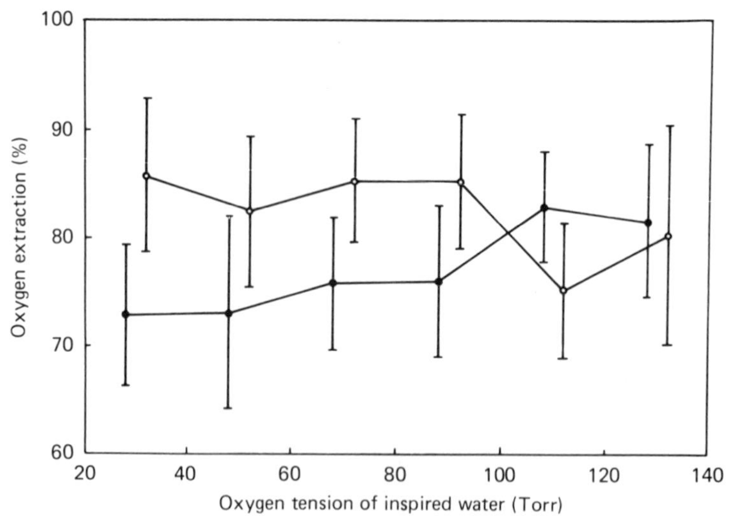

of a respiratory alkalosis caused by the hyperventilation, or, more prolonged, by alterations in the intra-red cell organic phosphates modulating the oxygen affinity (Lykkeboe & Weber, 1978; Johansen, 1980). The compensation results in a maintained or diminished reduction in oxygen uptake at reduced oxygen availability. Again using the example of the carp, a fish able to avoid reduction in the standard oxygen uptake from normoxia down to a water P_{O_2} of 25–30 Torr (Beamish, 1964), the gradient along which oxygen is transported at unchanged rates to the tissues changes from about 140 Torr during normoxia to 25–30 Torr in hypoxia.

Acclimation of gill function to hypoxia

For many fishes, gill gas exchange acclimation to hypoxia is apparent as an ability to have higher oxygen uptakes after a period of hypoxia exposure compared to acute hypoxia exposure of similar severity. As an example, will two weeks of hypoxia acclimation in carps result in a 15% higher oxygen extraction than for normoxic carps acutely exposed to hypoxia (Fig. 4)? Individual carps showed oxygen extractions at the amazing value of 96%. This allows maintenance of oxygen uptake at a lower ventilatory requirement after hypoxia acclimation. Indeed the acclimated carps show a lower ventilation than normoxia-acclimated carps during hypoxic exposure (Table 2). Beamish (1964) demonstrated a smaller cost of breathing in hypoxia-acclimated compared with nor-

Table 2. *Gill water flow* ($\dot{V}_g$) *and oxygen consumption* ($\dot{V}_{O_2}$) *at high and low inspired oxygen tension* (P_{iO_2}) *in normoxia- and hypoxia-acclimated carp. (From Lomholt & Johansen, 1979)*

	Normoxia-acclimated	Hypoxia-acclimated
$P_{iO_2} > 100$ Torr:		
$\dot{V}_g$ (ml kg^{-1} min^{-1})	195 ± 69[a]	212 ± 143
$\dot{V}_{O_2}$ (ml O_2 s.t.p. kg^{-1} h^{-1})	48.3 ± 11.9	49.1 ± 31.2
$P_{iO_2} < 40$ Torr:		
$\dot{V}_g$ (ml kg^{-1} min^{-1})	1122 ± 415	799 ± 294
$\dot{V}_{O_2}$ (ml O_2 s.t.p. kg^{-1} h^{-1})	62.0 ± 19.2	50.9 ± 20.9

[a] All measurements are average ± s.d. of five fish.

moxia-acclimated carps in hypoxic water. The increased potential for gill oxygen uptake after hypoxia acclimation is of obvious advantage by allowing higher scopes for needed physical activity in food and habitat selection.

The acclimation pattern may vary in different species. Thus Kerstens, Lomholt & Johansen (1979) demonstrated that, following hypoxia acclimation, the flounder, *Platichthys flesus,* had an oxygen uptake twice as large as it had during acute hypoxia exposure. Notably, however, this was associated with an unchanged oxygen extraction but a doubling of the gill ventilation (Fig. 5). In *P. flesus* the oxygen extraction from the ventilatory current averaged 76% in normoxic water falling to 52% during acute hypoxia and remaining at this level throughout an acclimation period of three weeks. The study by Kerstens *et al.* (1979) has special interest in that no physical disturbance was inflicted on the buried fish to obtain the continuous measurements of ventilation and exhaled water gas composition. Glass or plastic funnels were loosely impressed into the sand over the mouth and opercular openings allowing total ventilated water flow and exhaled water to be sampled from fish voluntarily buried in sand.

In another species, *Platichthys stellatus* (in the same weight range), Wood, McMahon & McDonald (1979) reported very similar values for oxygen extraction averaging 67.8% from well-aerated water at temperatures similar to those in

Fig. 5. Ventilation volume (open columns) and percentage extraction of oxygen from the ventilatory current in the flounder, *Platichthys flesus*. The columns are averages for each group. Bars express ±1 s.d. Student's *t*-test shows a significant difference between ventilation in normoxic versus acutely hypoxic ($P < 0.01$), normoxic versus chronically hypoxic ($P < 0.001$) and acutely versus chronically hypoxic ($P < 0.001$) flounders, and a significant difference in oxygen extraction between normoxic and acutely hypoxic ($P < 0.001$) as well as normoxic versus chronically hypoxic ($P < 0.001$) fish. Number of fish in each group: normoxia, $n = 7$; acute hypoxia, $n = 7$; chronic hypoxia, $n = 5$. (From Kerstens *et al.*, 1979.)

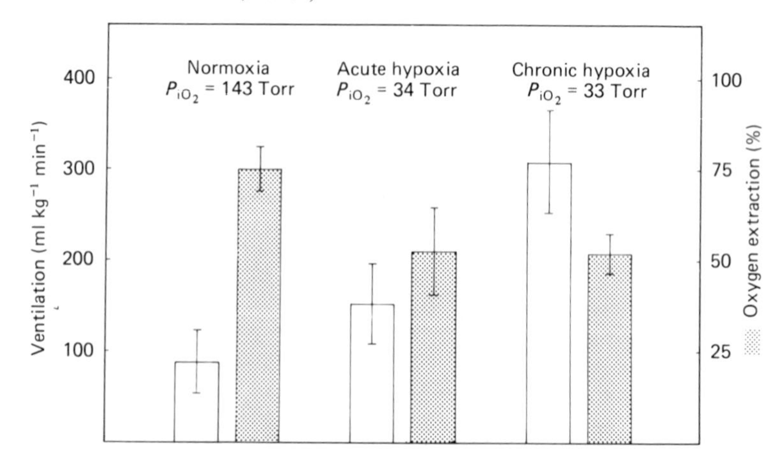

the study of Kerstens *et al.* (1979). The oxygen uptake rates were also closely similar for the species (0.45 ml O_2 kg^{-1} min^{-1} for *P. flesus* and 0.458 ml O_2 kg^{-1} min^{-1} for *P. stellatus*). The measured ventilation for *P. stellatus*, however, was higher than for *P. flesus* (109.2 ml kg^{-1} min^{-1} against 88.6 ml kg^{-1} min^{-1}) resulting in a much higher ventilatory requirement for *P. stellatus*. This could either reflect differences in experimental techniques, since the flounders studied by Wood *et al.* (1979) had rubber sleeves sutured around their opercular openings, or, alternatively, *P. stellatus* could have a lower gill transfer factor for oxygen (Table 3).

Which strategy of compensation different species of fish employ in compensating for hypoxia may relate to the cost of breathing and the capacity for alteration of lamellar surface area, ventilation, blood flow and its distribution.

How the fish gill vasculature responds to hypoxia

It appears unresolved whether fish gills possess a vascular bypass route of the secondary lamellae or not (see Randall, *Blood flow through gills,* in this volume). The consensus for teleost fish seems to be that with a few exceptions, e.g. the eel, *Anguilla anguilla* (Steen & Kruysse, 1964) and the Channel catfish, *Ictalurus ictalurus* (Boland & Olson, 1979) no discrete bypass vessels or channels exist (Laurent & Dunel, 1976). Blood flow distribution within secondary lamellae may, however, create conditions for gas exchange which result in a 'physiological' dead space or shunt effect. For elasmobranch fishes the situation seems more clear since approximately one-fifth of the total respiratory exchange area is located within a ventilatory dead-space area resulting in a respiratory bypass shunt. Also, in each gill filament there are non-respiratory capillaries arising from both the afferent and efferent side of the lamellar vasculature, allegedly serving a nutritive function (Cooke, 1980).

Accepting that no discrete bypass routes past the secondary lamellae exist in teleost gills, the well-known vasoconstrictor response of gills to hypoxia (Holeton & Randall, 1967) may be maladaptive to an oxygen-deficient ambient condition, unless the vasoconstriction diverts blood to larger and better-ventilated areas of the gas exchange surfaces or influences the diffusion distance or barrier in the lamellae. From fish head perfusion experiments (Pettersson & Johansen, 1981) it has indeed been concluded that this is the case. A myogenic vasoconstrictor response located in the efferent lamellar arterioles will, when stimulated by hypoxia, lead to an increased afferent lamellar pressure causing perfusion (recruitment) of unperfused lamellae, and hence increase the gill functional surface area. If increased adrenergic activity accompanies the hypoxic stimulus, which is likely to be the case, gill vasodilatation at other sites in the gill vasculature (efferent filament arterioles) will lower overall branchial vascular resistance which, in combination with the adrenergic effect on the heart, will increase arterial outflow from the gills. A lower P_{O_2} gradient between arterial blood and inspired water, occurring simultaneously, will optimise oxygen transfer during the hypoxic exposure. That gills in fishes resting in normoxic water are perfused less than maximally, i.e. distal lamellae are not perfused, has recently been demonstrated for rainbow trout (Booth, 1979). However, for species like the carp it is difficult to envisage how oxygen extractions exceeding 80% at rest can be possible without involvement of the maximum gill exchange area.

The gill transfer factor for oxygen (T_{O_2}), diffusing capacity (D_{gO_2}) and its dependence on oxygen uptake, gill ventilation and perfusion

The diffusive transfer of oxygen between ambient water and blood can conveniently be expressed by the ratio of oxygen uptake (V_{O_2}) and the mean P_{O_2} difference between the water passing the secondary lamellae and the blood flowing inside them $(T_{O_2})(D_{gO_2}) = [\dot{V}_{O_2}/(P_i + P_e)^{-2}_{O_2} - (\mathrm{P_a} + \mathrm{P_v})^{-2}_{O_2}]$. (The kinetics of oxygen binding to haemoglobin are disregarded in this expression.)

Using data for the dogfish, *Scyliorhinus*, Piiper, Meyer, Worth & Willmer (1977) have compared this arithmetic mean method both with a procedure based on an equation for gill counter-current exchange assuming a linear oxygen–haemoglobin dissociation curve (Scheid & Piiper, 1976), and with a method based on Bohr integration of the oxygen equilibrium curve for *Scyliorhinus* and the same counter-current model. Using these three methods on *Scyliorhinus* gave strikingly similar results. These results showed that the diffusing capacity P_{O_2} showed a tendency to increase during swimming when the oxygen uptake was increased. The swimming of *Scyliorhinus* was not very vigorous and similar data on trout swimming at far greater intensity showed the gill oxygen transfer factor to increase more than fourfold (Table 3).

The tuna fishes excel among the other species with an estimated gill oxygen transfer factor (diffusing capacity) of a magnitude exceeding that for all other fish studied (Stevens, 1972). The value in fact exceeds that for the most agile among the air-breathing poikilotherms, the varanid lizards, which have diffusing capacities of 0.075 ml O_2 kg^{-1} min^{-1} $Torr^{-1}$ at 37°C (Glass, Johansen & Abe, 1981).

The high oxygen transfer factor estimated for the skipjack tuna (Stevens, 1972) correlates with an oxygen uptake rate higher than that for other fish studied experimentally at similar temperatures (Brett, 1964). The gill surface areas of species of tuna also exceed those for other fish studied and approach the lung surface areas of mammals (Muir, 1969). As pointed out by Stevens (1972) the exceptionally high transfer factor for tuna fish gills suggests that ions and other diffusible matter traversing the gill exchange surfaces will also be exchanged at low resistance. Although an obvious advantage for gas exchange, high fluxes of ions may require corrective homeostatic measures.

Another important inference should be made when comparing the transfer factors of gas exchangers in vertebrates with the operation of counter-current exchange in fishes. One advantage of tidal ventilation in a lung is often stated to be the stability of the gaseous environment facing the alveolar surfaces in the lung. This stability will require that the oxygen removal from the lung (oxygen extraction) remains rather low and stable. In a counter-current fish gill much higher oxygen extractions are possible without jeopardising arterial P_{O_2} stability provided that the ambient water remains stable in oxygen content. The overall consequence may be that very low mean P_{O_2} gradients between water and blood prevail in fish and hence give higher apparent diffusing capacities than, for instance, values estimated from morphometric studies alone.

In another study on the elasmobranch *Squalus suckleyi,* the ventilation past

Table 3. *Oxygen transfer factors* (T_{O_2}) *or gill diffusing capacities* ($D_{g\,O_2}$) *for various species of fish*

Species	(T_{O_2}) ($D_{g\,O_2}$) (ml O_2 kg^{-1} min^{-1} $Torr^{-1}$)	Temperature (°C)	Reference
Salmo gairdneri (rainbow trout)			
Resting	0.0056	15	Randall *et al.* (1967)
Swimming	0.027	15	Randall *et al.* (1967)
Scyliorhinus stellaris			
(dogfish)	0.009	18	Piiper & Schumann (1967)
Resting	0.015	18	Piiper *et al.* (1977)
Moderately swimming	0.018	18	Piiper *et al.* (1977)
Squalus suckleyi (Pacific dogfish)	0.011	8–10	Hanson & Johansen (1970)
Katsowonus pelamis (tuna)	0.115	24	Stevens (1972)
Chaenocephalus aceratus (icefish)	0.0027	0	Holeton (1970)
Platichthys stellatus (starry flounder)	0.0069	10	Wood *et al.* (1979)
Ameiurus nebulosus (bullhead catfish)			
Normoxia	0.0068	24	Fisher *et al.* (1969)
Hypoxia ($P_{i\,O_2}$)	0.0108	24	Fisher *et al.* (1969)

the gills was varied artificially by placing the fish in a divided chamber allowing the trans-gill (buccal to opercular) hydrostatic gradient to be altered (Hanson & Johansen, 1970). An extended range of gill ventilations was obtained by curarising the fish and thus obtaining higher and lower values for ventilation than are possible in unanaesthetised fish. The calculated oxygen uptake and percentage extraction from the ventilatory current in these experiments are depicted in Fig. 6. These results show that oxygen uptake in *Squalus* is clearly ventilation limited within the normal ventilation range for intact fish (shaded area) and that oxygen extraction from the ventilatory current is reciprocally related to ventilation within the same range. This relation is typical for most fish but it should be remembered that other species, like the carp (see earlier), are able to maintain an unchanged oxygen extraction in the face of five-fold or higher ventilation increases.

If the gill transfer factor (diffusing capacity) for oxygen is calculated within

Fig. 6. Relation of ventilation to oxygen uptake (bottom) and oxygen extraction in normal (solid circles) and curarised (triangles) pacific dogfish, *Squalus suckleyi* (6–7°C). Open circles from the dogfish *Scyliorhinus stellaris* (16°C) (Baumgarten-Schumann & Piiper, 1968). Hatched background indicates normal ventilation range. Mean values ± 1 s.d. (From Hanson & Johansen, 1970.)

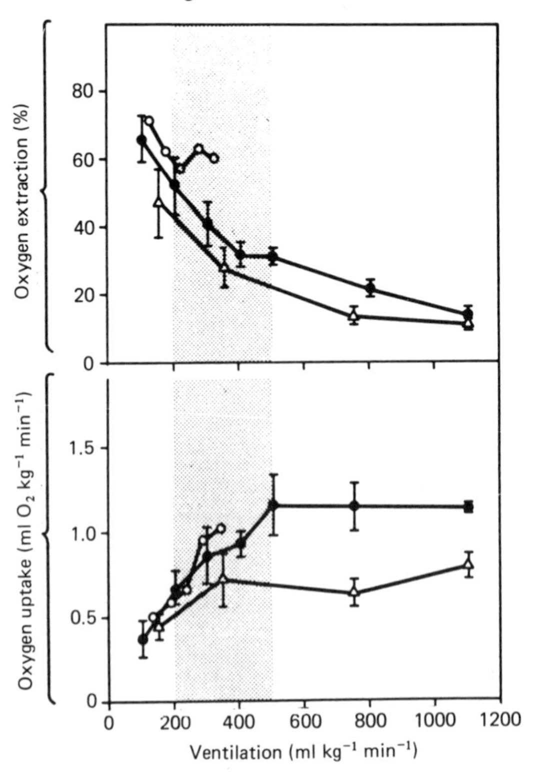

the range of rising oxygen uptake, a linear relation is obtained (Fig. 7). This implies that the effective mean trans-gill P_{O_2} gradients are maintained when the oxygen uptake rate varies by four-fold or more.

Piiper *et al.* (1977) working on *Scyliorhinus* also recorded a well-defined linear relation between gill oxygen uptake and respiratory water flow (ventilation) (Fig. 8). Moderate swimming activity, however, appeared to be associated with an increase in the ventilatory requirement ($\dot{V}_g/\dot{V}_{O_2}$).

Oxygen uptake in elasmobranchs is also clearly perfusion dependent, this being apparent from both the *Scyliorhinus* (Piiper *et al.*, 1977) and *Squalus* (Hanson & Johansen, 1970) studies. Fig. 9 documents this for *Squalus*. Swimming activity in *Scyliorhinus* increases the relative requirement for gill perfusion ($\dot{Q}/\dot{V}_{O_2}$). For the sturgeon, *Acipenser transmontanus,* Burggren & Randall (1978) have demonstrated a linear relation between oxygen uptake and dorsal aortic P_{O_2} which indicates that oxygen uptake is diffusion limited and thus dependent upon the P_{O_2} gradient from arterial blood to mitochondria.

With gill oxygen uptake depending on both the ventilation and circulation, the ratio between the two convective flows, the ventilation–perfusion ratio, will predictably have an optimal range for efficient gill function. Fig. 10 shows how the oxygen tension in arterial blood and exhaled water is related to the ventilation–perfusion ratio in *Squalus*. Oxygen extraction and the efficacy of oxygen removal from water fall as the ventilation–perfusion ratio increases. In resting *Squalus* normal values for the ratio ranged widely between 6 and 25, but were typically between 9 and 15. Baumgarten-Schumann & Piiper (1968) reported a ratio of 9

Fig. 7. Diffusing capacity (D_{gO_2}) for gills of *Squalus suckleyi* related to gill oxygen uptake at 6° and 10°C.

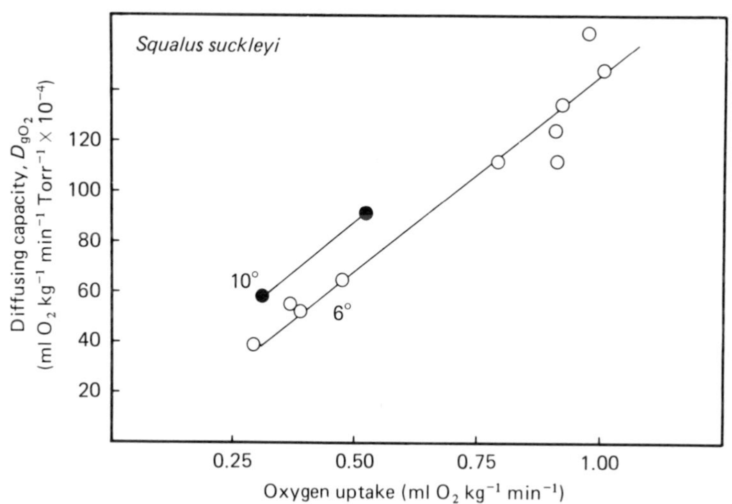

Fig. 8. Relation between respiratory water flow (ventilation) and oxygen uptake during rest (solid circles) and spontaneous swimming (open circles). (From Piiper *et al.*, 1977.)

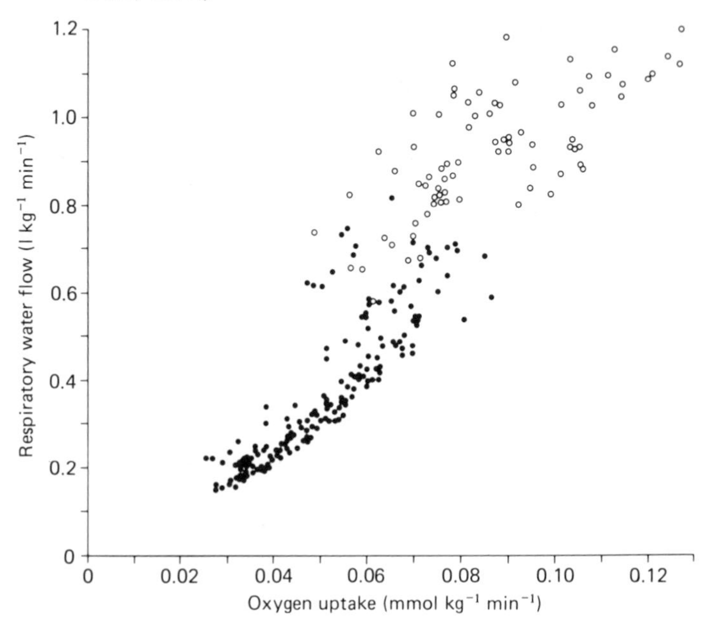

Fig. 9. Increase in oxygen uptake in *Squalus* with increased cardiac output, estimated by the Fick principle (mean values ±1 s.d.). (From Hanson & Johansen, 1970.)

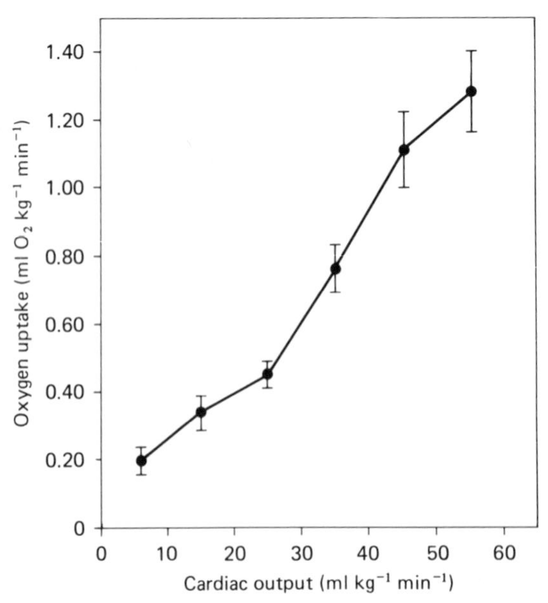

for *Scyliorhinus stellaris* at a ventilation of 200 ml kg^{-1} min^{-1}, while Robin, Murdaugh & Millen (1964) estimated a ratio of 18 at ventilation levels of about 450 ml kg^{-1} min^{-1} for *Squalus acanthias*. Fig. 10 for *Squalus suckleyi* shows that the P_{O_2} gradient from arterial blood to exhaled water is positive below ventilation–perfusion ratios of about 16, a finding which suggests that the efficiency of gas exchange may be higher at this or lower levels. Positive arterial to expired water P_{O_2} gradients signify the operation of an apparent counter-current gas exchange between blood and water in the gills. Although such exchange is clearly operational in teleost gills (see the chapter by Piiper & Scheid, *Physical principles of respiratory gas exchange in fish gills,* in this volume), it has been questioned for elasmobranch gills due to their different gill structure which involves an interbranchial septum. Grigg (1970) and Grigg & Read (1971) contested this, however, and offered evidence that ventilated water in elasmobranchs indeed flows unidirectionally between the secondary lamellae before entering the septal canals, thus leaving no structural hindrance for an operational counter-current exchange.

Fig. 10. Relations between (*a*) the ventilation–perfusion ratio ($\dot{V}_g/\dot{Q}_T$) and both the arterial P_{O_2} (closed circles) and exhaled water P_{O_2} (open circles) and (*b*) the ventilation–perfusion ratio and oxygen extraction from the ventilatory current in *Squalus suckleyi*. (From Hanson & Johansen, 1970.)

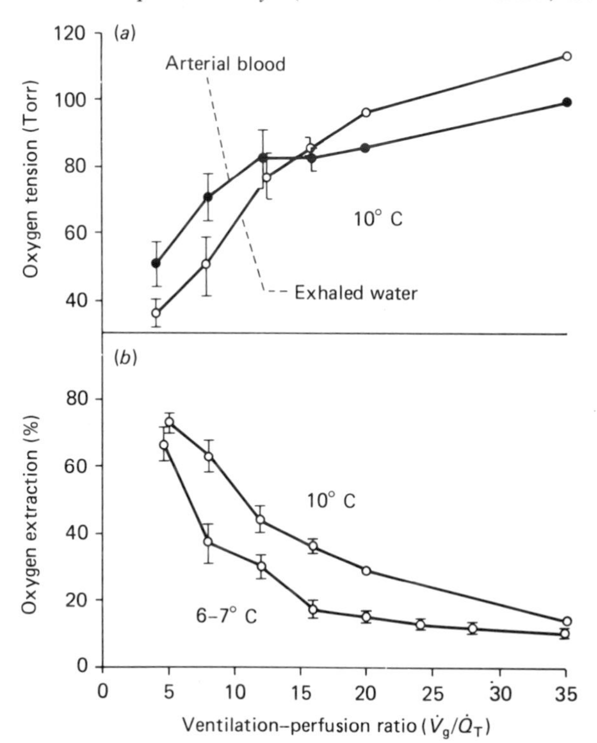

The results of Piiper *et al.* (1977) on *Scyliorhinus* also document that a counter-current exchange is operational at high levels of oxygen extraction (Fig. 11). Interestingly, oxygen extraction declines during swimming and the counter-current exchange is abolished or masked (Fig. 11). This may be related to the finer matching of water and blood at the lamellar level, perhaps caused by gill derangement during ram ventilation of swimming. The overall ventilation–perfusion ratio changed insignificantly from about 11.2 to 10 going from rest to moderate swimming.

The gill transfer factors for oxygen among the teleost species vary greatly (Table 3). The value calculated by Wood *et al.* (1979) for the starry flounder, *P. stellatus,* may not only relate to the rather sluggish nature of the species, but may also be a consequence of the very high blood affinity for oxygen in the species. The authors argue that a high oxygen affinity will ensure a high arterial saturation at low arterial P_{O_2} and thus allow the species to transport oxygen to the tissues at low arterial tensions. This enables the fish to maintain a much higher mean P_{O_2} gradient across the gill exchange surfaces without sacrificing arterial saturation. High trans-gill P_{O_2} gradients would in turn imply lower values for the gill oxygen transfer factor (diffusing capacity) (Table 3). This argument raises a question which has been posed earlier and answered affirmatively, in the context of a comparison between blood oxygen affinity and circulating arterial

Fig. 11. Expired–arterial P_{O_2} difference, $(P_e - P_a)_{O_2}$, plotted against expired P_{O_2} during rest (filled symbols) and swimming (open symbols) for the dogfish, *Scyliorhinus stellaris*. (From Piiper *et al.*, 1977.)

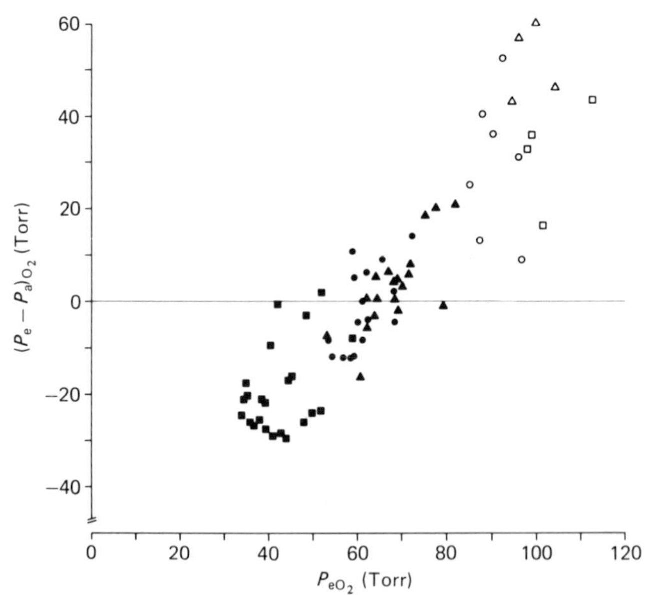

P_{O_2}, in a series of amphibians having wide differences in oxygen affinity (Lenfant & Johansen, 1967). Are the ventilation–perfusion ratio and diffusion characteristics of the gas exchange surfaces of an animal 'tuned' to result in the circulating arterial P_{O_2}s matching the oxygen affinity of the haemoglobin of the species in order to attain near maximal oxygen saturations? There appear to be too few data available to answer this question yet for fishes.

Interestingly, for the Antarctic haemoglobin-free icefish, *Chaenocephalus aceratus*, the oxygen capacity of the blood equals that of saline, which markedly elevates the demand for blood perfusion in oxygen delivery. This brings the ventilation–perfusion ratio closer to unity compared with values of around ten or higher typical of fish. In spite of low aerobic oxygen requirements, the ventilatory and circulatory convection requirements become higher than those typical for fishes (Table 1).

All the results on gill oxygen transfer (diffusing capacity) in Table 3 have been calculated on the basis of the arithmetic mean method for the trans-lamellar P_{O_2} difference, except those for the bullhead catfish. For this species the carbon-monoxide diffusing capacity was experimentally measured and the diffusing capacity for oxygen was calculated from that value. The authors claimed that no significant correlation existed between D_{gCO} and the ventilation or oxygen uptake via the gills. This is contrary to the results for *Squalus suckleyi* quoted above which were obtained using the arithmetic mean method for the trans-gill P_{O_2} gradient.

The energy requirements of fish gills

The many functions which gills have in addition to respiratory gas exchange are likely to require energy. How gill tissues satisfy their oxygen requirements has never been properly investigated, presumably because the difficulties in separating the gas exchanger function of the gills from the gas exchange requirements of gill tissues proper have been difficult or insurmountable. Some recent simple perfusion experiments have offered suggestive evidence on this important point (Johansen & Pettersson, 1981). Both single gill arches and whole head preparations freshly excised from the Atlantic cod, *Gadus morhua*, were studied.

Fig. 12 shows the perfusion arrangement of a single gill arch. The basis for the method is that the perfusion fluid, saline, is similar in composition to the fluid surrounding the gill arch. This fluid is held in a closed chamber and serves as a fluid reservoir for the gill perfusion. In this way no gradient in P_{O_2} can be set up between the perfusion fluid entering the afferent filament artery and the fluid surrounding the gill arch preparation, and the gas exchanger function of the gill arch is eliminated. Owing to the structure of gills, many cells which build gill tissues will be at similar diffusion distances from water and blood in intact fish. An appropriate question must be 'what fraction of the total gill oxygen requirement may be supplied by direct diffusion from the outside without the intervention of blood circulating internally?'. From Fig. 12 it becomes apparent that when the chamber is closed, oxygen-electrode A will allow calculation of the oxygen consumption of the entire gill arch system. The difference between the readings of oxygen-electrodes A and B (the latter samples the effluent perfusate from the gill arch) and information on the flow rate of the pump will offer information about how much oxygen is removed from the perfusate to the gill tissues facing the internal perfusion path of the system. The difference between the total oxygen uptake, based on readings from electrode A minus the oxygen removed during the internal passage of the perfusate, will indicate what fraction of the total oxygen uptake was satisfied by direct diffusion to metabolising tissues from the ambient medium independently of the internal gill perfusion. The removal of oxygen during the internal passage will admittedly set up a P_{O_2} gradient between outside and inside which increases towards the outflow end of the gill arch preparation. This will cause some inward movement of oxygen molecules not directly involved in metabolism and will thus cause an overestimation of the fraction of oxygen taken up directly from the ambient water. With this limitation, and the obvious one related to the gill tissues not being uniformly

Fig. 12. Gill arch perfusion in a closed respirometer for determination of gill tissue oxygen consumption. See text for further explanation. (From Johansen & Pettersson, 1981.)

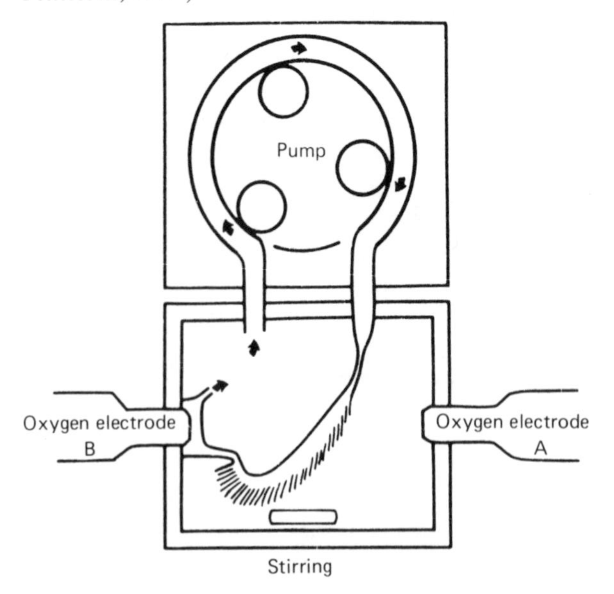

exposed to saline both outside and in, the results nevertheless afford a first insight into the aerobic metabolic requirements of intact gills (Table 4). In series A, the saline bathing the perfused arch was initially kept aerated in order to determine the internal oxygen uptake at stable conditions while later the chamber was closed to allow assessment of the total oxygen requirement.

The gill oxygen uptake from the inside (the perfusate) was almost the same in the two series. The fraction of total oxygen uptake 'satisfied' by direct diffusion from the outside was as high as 41.7% in series A compared to 26.7% in series B. If these values for total and fractional oxygen uptake by gills are compared with the oxygen requirements of intact cod, it can be seen that the gills may be responsible for 6.64% of the total fish oxygen uptake. This gives gill tissue a weight-specific oxygen uptake of 95 $\mu l\ O_2\ g^{-1}\ h^{-1}$, a value nearly twice that of intact fish. Approximately 3% of the total oxygen uptake by the fish is acquired and consumed directly by diffusion into gill tissue from ambient water (Johansen & Pettersson, 1981).

Gill tissue homogenates, which will never reflect true oxygen uptakes of intact tissues, showed high values of oxygen uptake exceeding those of liver and heart muscle and approaching those of kidney tissues. (H. Gesser & K. Johansen, unpublished observations, 1981.)

These results document that fish gills have an important oxygen requirement and that about half of this is satisfied by direct diffusion from the ambient water. This component and the considerable contribution to total oxygen uptake by cutaneous gas exchange reduce the perfusion requirement of fish gills. This implies that a correct employment of the Fick principle for cardiac output calculation in fish should be based on oxygen uptake values derived from measurement of gill ventilation and oxygen extraction corrected for the oxygen fraction taken up by gill tissues directly from the ambient water.

Table 4. *A comparison of oxygen uptake from the perfusion fluid ('internal') and direct from the ambient medium ('external') of isolated gill arches (g gw: gram gill weight). (From Johansen & Pettersson, 1981)*

	Saline P_{O_2} (Torr)	Total gill oxygen uptake ($\mu l\ O_2$ (g gw)$^{-1}$ h^{-1})	'Internal' oxygen uptake ($\mu l\ O_2$ (g gw)$^{-1}$ h^{-1})	'Internal' (%)	'External' (%)	*n*
Series A	150	95.3 ± 7.0	57.0 ± 8.9	58.3 ± 6.8	41.7 ± 6.8	9
Series B	100	76.3 ± 12.8	55.7 ± 10.4	73.7 ± 7.5	26.7 ± 7.5	5

Air-breathing fishes

For the many air-breathing fishes among teleosts and holosteans the gills have a reduced importance in gas exchange and also reduced anatomical development in the adult stages. The treatment of this subject has been thoroughly covered in recent reviews (Johansen, 1970; Randall, Burggren, Farrell & Haswell, 1981) and will not be treated extensively here. For the so-called facultative air-breathing fish, in which the gills are normally well-developed and suffice to support the entire gas exchange requirement in well-aerated water, the situation is somewhat different. The bowfin, *Amia calva*, is a case in point in that its dependence on gill gas exchange and/or air breathing depends on the water temperature, or more directly on the metabolic requirement related to the body temperature of the fish (Johansen, Lenfant & Hanson, 1970). Fig. 13 shows how the gills and air bladder share in the total gas exchange as the fish body temperature changes between 10 and 30°C; whereas oxygen uptake at 10°C is almost exclusively covered by gill gas exchange, at 30°C the air bladder has

Fig. 13. Contributions from the gills and air bladder to total gas exchange in relation to ambient water temperature for the bowfin, *Amia calva*. (Data from Johansen *et al.*, 1970.)

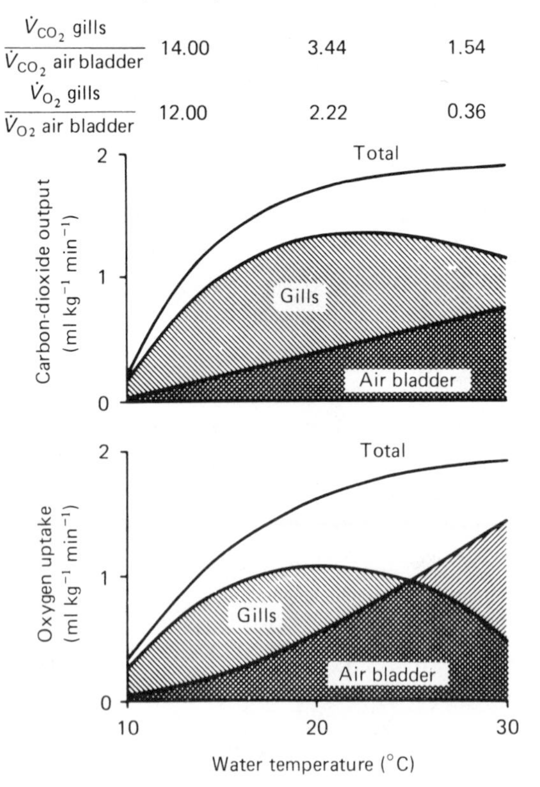

assumed more than two-thirds of the total oxygen uptake. For carbon dioxide exchange the gills are seen to play a more conservative role at all temperatures, being the dominant exit route also at 30°C. Interestingly, between 5° and about 25°C water temperature, when air breathing assumes major importance in oxygen uptake, the cardiac output rises five to six times showing total oxygen uptake to be dependent upon the overall rate of perfusion. Notably, however, as the water temperature increases and the air bladder becomes the predominant oxygen uptake organ, the vascular resistance of the gills proper falls rather drastically to one-third of its value at 5°C when water temperature has reached 25°C. This disparate relation between total blood flow through the branchial region and the gill vascular resistance offers clear evidence of highly vasoactive pathways within the branchial vasculature.

Lungfishes

For all three genera of lungfishes save the Australian *Neoceratodus forsteri*, the lung is the principal gas exchange organ in the adult forms. All species of lungfish larvae have, however, external gills of an alleged, but not yet quantified, importance in gas exchange. It has been demonstrated that when lungfish larvae are reared in well-oxygenated water the size of the external gills diminishes, while conversely, exposure to hypoxic water causes the external gills to greatly hypertrophy.

The gill-like structure on the posterior appendages of the male *Lepidosiren*, which is the form of a very vascular tuft-like structure, was deduced by Kerr (1898) to have an important role in parental rearing of the *Lepidosiren* larvae. Two suggestions for the functional significance of this peculiar structure have been offered; one being that it functions as a gill for the adult male, allowing the fish to satisfy its oxygen requirements from water without having to leave the larvae for excursions to the surface for air breathing. The other implies that the structure is used to transfer oxygen from the atmosphere, obtained by the adult air breathing at the surface, to the immediate vicinity of the larvae by release of absorbed oxygen via the fin gill-like structures. Experimental proof or verification of these interesting ideas is much needed.

For post-larval development of young *Protopterus amphibius*, it has been clearly demonstrated that the fraction of the total oxygen requirement derived from water via the skin and remaining external gills declines with growth of the fish (Johansen, Lomholt & Maloiy, 1976).

Fig. 14 shows that among adult lungfishes the Australian *Neoceratodus* depends exclusively on its gills for total gas exchange in well-aerated water, while species of *Protopterus* and *Lepidosiren* are obligate air breathers and actually drown if denied access to the atmosphere (Johansen, 1970). If we compare the performance of *Neoceratodus* gills with those of teleosts (Table 1) we find values comparable to those of species with the rather sluggish habits which characterise *Neoceratodus*.

Gill function in amphibia

In larval amphibians external and/or internal gills must, together with the general body surface, play a major role in respiratory gas exchange. Except for the external gills of neotenic adult urodeles, amphibian gills can be viewed as temporary or disposable gas exchange organs but have received surprisingly little interest from physiologists. Anatomists, however, have often investigated problems related to how the vascular supply to the gills in larvae is transformed or accommodated into the permanent vascular arrangement associated with lung breathing in adult forms.

For anuran amphibians discrete vascular shunts between the afferent and efferent gill arteries have recently been described both for *Rana temporaria* and *Bufo bufo* (de Saint-Aubain, 1981). The shunts are alleged to permit continued circulation after metamorphosis and atrophy of the external gills.

Many aquatic salamanders have trimodal gas exchange in which pulmonary, branchial and cutaneous surfaces may contribute to gas exchange. Guimond & Hutchison (1972) actually measured the fractional importance of lungs, gills and

Fig. 14. Relative contributions of aquatic and aerial breathing to total gas exchange in three species of lungfishes. Temperature: 20°C. (From Lenfant, Johansen & Hanson, 1970.)

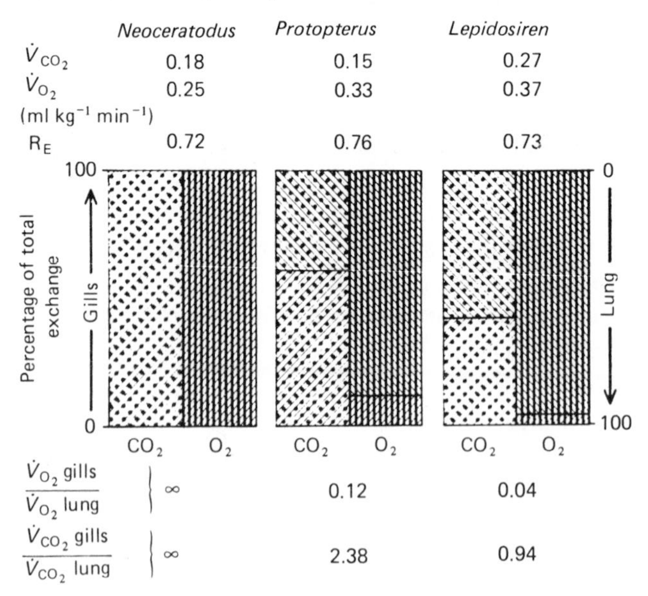

skin to total gas exchange in *Siren* and concluded that the gills accounted for only 2.5% of total oxygen uptake at 25°C. The gills showed a greater participation in carbon dioxide exchange amounting to 12% of the total at 25°C.

The only salamander for which the gills have a demonstrated great importance is the mud-puppy, *Necturus maculosus*. Based on results obtained using respirometry, partitioning of the lung and skin and gill gas exchange, it was demonstrated that the external gills contributed 53.6% of the total oxygen exchange and 45.6% of the total carbon dioxide exchange at 5°C, increasing to 60.2% and 61.3% respectively at 25°C (Guimond & Hutchison, 1972). The waving motion of the external gills, which obviously aided convective movement of the ambient water, was clearly stimulated both by increased water temperature and reduced oxygen content of the water.

The question of a reciprocal interaction between pulmonary and branchial gas exchange in salamanders received a most significant contribution from Figge (1936). Although his primary interest was in the problem of gill reduction in amphibian metamorphosis, his studies must clearly be regarded as pioneering in our understanding of how amphibian gill vessels react to external (aquatic) and internal (blood-borne) stimuli.

In contrast to the ongoing discussion about vascular bypass vessels passing the secondary lamellae in fish gills, there can be no doubt that gill bypass vessels are prominently present in urodele amphibians (Figge, 1936; Baker, 1949). Fig. 15, redrawn from Figge (1936), shows a prominent anastomotic vessel between the ventral and dorsal segments of an aortic arch allowing bypass of the external gill in *Amblystoma tigrinum*. Noteworthy is the separation of the large shunt vessel into what Figge (1936) termed anastomotic arterioles where these branch off from the main afferent branchial artery.

Most revealingly, Figge perfused the external gill system from the dorsal aortic parent vessel to the external gill. By moving the perfusion cannula past the origin of the shunt vessel, he could perfuse the external gill vessels without the anastomotic connection. Conversely, the shunt vessel could be perfused discretely by ligating the continuation of the afferent branchial vessels into the capillary sections of the external gill. Some of his results are reproduced in Fig. 15. Adrenaline added to the perfusion fluid caused a doubling of the perfusion rate through the external gill, while the shunt vessels clearly constricted to the same stimulus. Artificial ventilation of the lungs of *Amblystoma* with carbon dioxide induced a differential reaction similar to that produced by adrenaline, i.e. the shunt vessels constricted while flow through the external gill was promoted by dilatation. Ventilation of the lungs with oxygen caused a reaction opposite to that produced by carbon dioxide. These results are most important in showing that the control of lung function is closely interrelated with that of external gill function for the optimisation of gas exchange in a bimodal system. A recent study (Melvin, Dail & Wood, 1982) has demonstrated gill blood flow to be regulated in part by adrenergic innervation of the shunt vessels and by circulating catecholamines primarily acting on α- and β-adrenal receptors in the respiratory section of the external gills of *Amblystoma tigrinum*.

Fig. 15. (*a*) Schematic drawing of the vascular connections to an external gill in *Amblystoma tigrinum*. Note the discrete shunt vessel (anastomosis) between the afferent and efferent branchial vessels. Dashed line shows the position of a ligature which allows shunt vessel perfusion. (*b*) Effects of added adrenaline and oxygen content of the perfusion fluid on gill and shunt blood flow in *Amblystoma tigrinum*. (From Figge, 1936.)

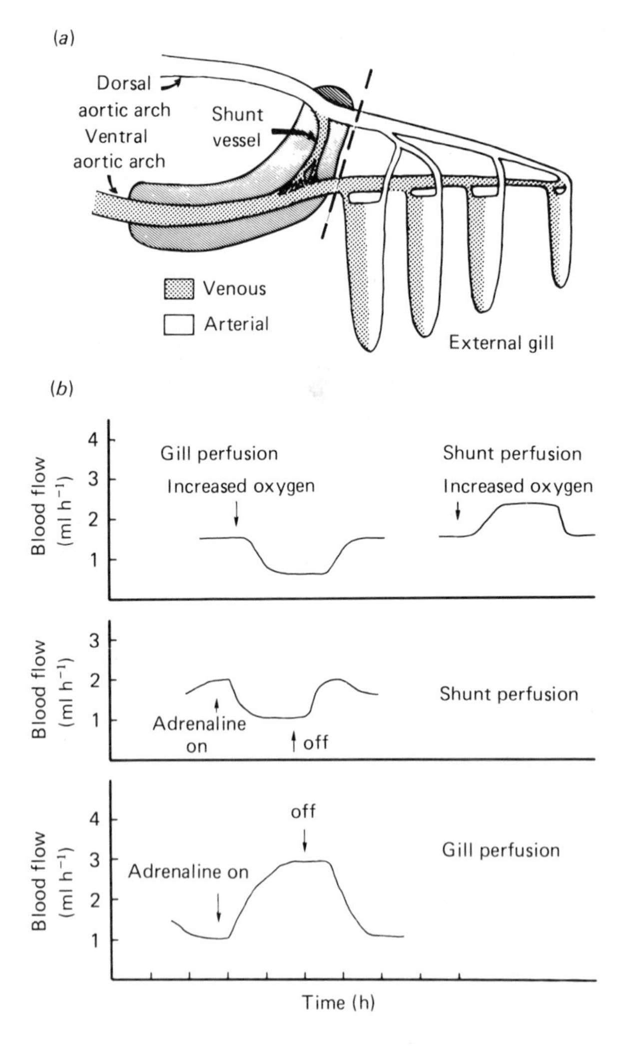

References

Baker, C. L. (1949). The comparative anatomy of the aortic arches of the urodeles and their relation to respiration and degree of metamorphosis. *Journal of the Tennessee Academy of Sciences,* **24,** 12–40.

Baumgarten-Schumann, D. & Piiper, J. (1968). Gas exchange in the gills of resting unanaesthetized dogfish (*Scyliorhinus stellaris*). *Respiration Physiology,* **5,** 317–25.

Beamish, F. W. H. (1964). Respiration of fishes with special emphasis on standard O_2 consumption. IV. Influence of carbon dioxide and oxygen. *Canadian Journal of Zoology,* **42,** 847–56.

Billet, F. S. & Courtenay, T. H. (1973). A stereoscan study of the origin of ciliated cells in the embryonic epidermis of *Amblystoma mexicanum. Journal of Embryology and Experimental Morphology,* **29,** 549–58.

Boland, E. J. & Olson, K. R. (1979). Vascular organization of the catfish gill filament. *Cell and Tissue Research,* **198,** 487–500.

Booth, J. H. (1979). The effect of oxygen supply, epinephrine and acetylcholine on the distribution of blood flow in trout gills. *Journal of Experimental Biology,* **83,** 31–9.

Brett, J. R. (1964). The respiratory metabolism and swimming performance of young sockeye salmon. *Journal of the Fisheries Research Board of Canada,* **21,** 1183–226.

Budgett, J. S. (1901). On the breeding habits of some West African fishes, with an account of the external features in the development of *Protopterus annecteus,* and a description of the larvae of *Polypterus lapradei. Transactions of the Zoological Society of London,* **16,** 115–36.

Burggren, W. W. & Randall, D. J. (1978). Oxygen uptake and transport during hypoxic exposure in the sturgeon, *Acipenser transmontanus. Respiration Physiology,* **34,** 171–83.

Cooke, I. R. C. (1980). Functional aspects of the morphology and vascular anatomy of the gills of the endeavour dogfish, *Centrophorus scalpratus* (McCulloch) (Elasmobranchii: Squalidae). *Zoomorphologie,* **94,** 167–83.

Dejours, P., Armand, J. & Verriest, G. (1968). Carbon dioxide dissociation curves of water and gas exchange of water breathers. *Respiration Physiology,* **5,** 23–33.

Figge, F. H. J. (1936). The differential reaction of the blood vessels of a branchial arch of *Amblystoma tigrinum* (Colorado axolotl). I. The reaction to adrenaline, oxygen, and carbon dioxide. *Physiological Zoology,* **9,** 79–101.

Fisher, T. R., Coburn, R. F. & Forster, R. E. (1969). Carbon monoxide diffusing capacity in the bullhead catfish. *Journal of Applied Physiology,* **26,** 161–9.

Freadman, M. A. (1981). Swimming energetics of striped bass (*Morone saxatilis*) and bluefish (*Pomatomus saltatrix*): hydrodynamic correlates of locomotion and gill ventilation. *Journal of Experimental Biology,* **90,** 253–65.

Garey, W. F. (1967). Gas exchange, cardiac output and blood pressure in free-swimming carp (*Cyprinus carpio*). Ph.D. Dissertation, State University of New York at Buffalo.

Gerald, J. W. & Cech, J. J., Jr. (1970). Respiratory responses of juvenile catfish (*Ictalurus punctatus*) to hypoxic conditions. *Physiological Zoology,* **43,** 47–54.

Glass, M. L., Johansen, K. & Abe, A. S. (1981). Pulmonary diffusing capacity in reptiles. (Relations to temperature and O_2 uptake.) *Journal of Comparative Physiology,* **142,** 509–14.

Greenwood, P. H. (1958). Reproduction in the East African lungfish, *Protopterus aethiopicus*, Heckel. *Proceedings of the Zoological Society of London*, **130,** 547–67.
Grigg, G. C. (1970). Water flow through the gills of Port Jackson sharks. *Journal of Experimental Biology*, **52,** 565–8.
Grigg, G. C. & Read, J. (1971). Gill function in an elasmobranch. *Zeitschrift für Vergleichende Physiologie*, **73,** 439–51.
Guimond, R. W. & Hutchison, V. H. (1972). Pulmonary, branchial and cutaneous gas exchange in the mud puppy, *Necturus maculosus maculosus* (Rafinesque). *Comparative Biochemistry and Physiology*, **42A,** 367–92.
Hanson, D. & Johansen, K. (1970). Relationship between gill ventilation and perfusion in the elasmobranch, *Squalus suckleyi*. *Journal of Fisheries Research Board of Canada*, **27,** 551–64.
Hemmingsen, E. A. & Douglas, E. L. (1970). Respiratory characteristics of the haemoglobin free fish, *Chaenocephalus aceratus*. *Comparative Biochemistry and Physiology*, **33,** 733–44.
Holeton, G. F. (1970). Oxygen uptake and circulation by a haemoglobinless Antarctic fish (*Chaenocephalus aceratus* Lonnberg) compared with three red-blooded Antarctic fish. *Comparative Biochemistry and Physiology*, **34,** 457–72.
Holeton, G. F. & Randall, D. J. (1967). Changes in blood pressure in the rainbow trout during hypoxia. *Journal of Experimental Biology*, **46,** 297–305.
Hughes, G. M. & Umezawa, S. (1968). On respiration in the dragonet *Callionymus lyra* L. *Journal of Experimental Biology*, **49,** 565–82.
Johansen, K. (1970). Physiology of Fishes II. In *Air breathing Fishes*, Vol. IV, ed. W. S. Hoar & D. J. Randall, pp. 361–411. London: Academic Press.
Johansen, K. (1980). Circulation of blood in vertebrate oxygen transport. In *Animals and Environmental Fitness*, ed. R. Gilles, pp. 133–55. Oxford: Pergamon Press.
Johansen, K., Lenfant, C. & Grigg, G. C. (1967). Respiratory control in the lungfish, *Neoceratodus forsteri* (Krefft). *Comparative Biochemistry and Physiology*, **20,** 835–54.
Johansen, K., Lenfant, C. & Hanson, D. (1970). Respiration in a primitive air breather, *Amia calva*. *Respiration Physiology*, **9,** 162–74.
Johansen, K., Lomholt, J. P. & Maloiy, G. M. O. (1976). Importance of air and water breathing in relation to size of the African lungfish *Protopterus amphibius*, Peters. *Journal of Experimental Biology*, **65,** 395–9.
Johansen, K. & Pettersson, K. (1981). Gill O_2 consumption in a teleost fish, *Gadus morhua*. *Respiration Physiology*, **44,** 277–84.
Kerr, R. J. (1898). The dry season habits of Lepidosiren. *Proceedings of the Zoological Society of London*, 41–4.
Kerstens, A., Lomholt, J. P. & Johansen, K. (1979). The ventilation, extraction and uptake of oxygen in undisturbed flounders, *Platichthys flesus*: responses to hypoxia acclimation. *Journal of Experimental Biology*, **83,** 169–79.
Kirsch, R. & Nonnotte, G. (1977). Cutaneous respiration in three freshwater teleosts. *Respiration Physiology*, **29,** 339–54.
Krogh, A. (1904). Some experiments on the cutaneous respiration of vertebrate animals. *Scandinavian Archives of Physiology*, **16,** 348–57.
Laurent, P. & Dunel, S. (1976). Functional organization of the teleost gill. I. Blood pathways. *Acta Zoologica* (Stockholm), **57,** 189–209.

Lenfant, C. & Johansen, K. (1967). Respiratory adaptations in selected amphibians. *Respiration Physiology,* **2,** 216–68.

Lenfant, C., Johansen, K. & Hanson, D. (1970). Bimodal gas exchange and ventilation–perfusion relationship in lower vertebrates. *Federation Proceedings,* **29,** 1124–9.

Liem, K. F. (1981). Larvae of air-breathing fishes as counter current flow devices in hypoxic environments. *Science,* **211,** 1177–9.

Lomholt, J. P. & Johansen, K. (1979). Hypoxia acclimation in carp – How it affects O_2 uptake, ventilation, and O_2 extraction from water. *Physiological Zoology,* **52,** 38–49.

Lykkeboe, G. & Weber, R. E. (1978). Changes in the respiratory properties of the blood in the carp, *Cyprinus carpio,* induced by diurnal variation in ambient oxygen tension. *Journal of Comparative Physiology,* **128,** 117–25.

Melvin, G., Dail, W. G. & Wood, S. C. (1982). Vascular adrenal receptors and adrenergic innervation in the gills of the salamander *Amblystoma tigrinum. Physiologist.* In press.

Muir, B. S. (1969). Gill dimensions as a function of fish size. *Journal of Fisheries Research Board of Canada,* **26,** 165–70.

Muir, B. S. & Kendall, J. I. (1968). Structural modifications in the gills of tunas and some other oceanic fishes. *Copeia,* 1968, 388–98.

Nonnotte, G. & Kirsch, R. (1978). Cutaneous respiration in seven seawater teleosts. *Respiration Physiology,* **35,** 111–18.

Pettersson, K. & Johansen, K. (1981). Hypoxic vasoconstriction and increased gas exchange efficiency by adrenaline in fish gills. *Journal of Experimental Biology,* in press.

Piiper, J., Meyer, M., Worth, H. & Willmer, H. (1977). Respiration and circulation during swimming activity in the dogfish, *Scyliorhinus stellaris. Respiration Physiology,* **30,** 221–39.

Piiper, J. & Schumann, D. (1967). Efficiency of O_2 exchange in the gills of the dogfish, *Scyliorhinus stellaris. Respiration Physiology,* **2,** 135–48.

Randall, D. J., Burggren, W. W., Farrell, A. P. & Haswell, M. S. (1981). *The Evolution of Air Breathing in Vertebrates.* 133 pp. Cambridge University Press.

Randall, D. J., Holeton, G. F. & Stevens, E. D. (1967). The exchange of oxygen and carbon dioxide across the gills of rainbow trout. *Journal of Experimental Biology,* **6,** 339–48.

Robin, E. D., Murdaugh, H. V. & Millen, J. E. (1964). Gill gas exchange in the dogfish shark. *Federation Proceedings,* **23,** 469.

Saint-Aubain, M. L. de (1981). Shunts in the gill filaments in tadpoles of *Rana temporaria* and *Bufo bufo* (*Amphibia, Anura*). *Journal of Experimental Zoology,* **217,** 143–6.

Saunders, R. L. (1962). The irrigation of the gills in fishes. II. Efficiency of oxygen uptake in relation to respiratory flow, activity and concentrations of oxygen and carbon dioxide. *Canadian Journal of Zoology,* **40,** 817–62.

Scheid, P. & Piiper, J. (1976). Quantitative functional analysis of branchial gas transfer: theory and application to *Scyliorhinus stellaris* (Elasmobranchii). In *Respiration in Amphibious Vertebrates,* ed. G. M. Hughes, pp. 17–38. London, New York: Academic Press.

Steen, J. B. and Kruysse, A. (1964). The respiratory function of the teleostean gills. *Comparative Biochemistry and Physiology,* **12,** 127–42.

Steffensen, J. F., Lomholt, J. P. & Johansen, K. (1981). The relative importance of skin oxygen uptake in the naturally buried plaice, *Pleuronectes platessa,* exposed to graded hypoxia. *Respiration Physiology,* **44,** 269–76.

Stevens, E. D. (1972). Some aspects of gas exchange in tuna. *Journal of Experimental Biology,* **56,** 809–23.

Watters, K. W., Jr. & Smith, L. S. (1973). Respiratory dynamics of the starry flounder *Platichthys stellatus* in response to low oxygen and high temperature. *Marine Biology,* **19,** 133–48.

Whiting, H. P. & Bone, Q. (1980). Ciliary cells in the epidermis of the larval Australian dipnoan, *Neoceratodus. Journal of the Linnean Society of London,* **68,** 125–37.

Wood, C. M., McMahon, B. R. & McDonald, D. G. (1979). Respiratory gas exchange in the resting starry flounder, *Platichthys stellatus*: a comparison with other teleosts. *Journal of Experimental Biology,* **78,** 167–79.

A.P.M.LOCKWOOD, S.R.L.BOLT &
M.E.DAWSON

Water exchange across crustacean gills

Within the Crustacea four different stratagems for the regulation of the body fluids and cell contents can be observed when the medium concentration is changed (Fig. 1). (1) Osmoconforming forms, such as the spider crab *Maia*, which rarely experience salinity change in their natural, sub-littoral environment, show little capacity to regulate the blood concentration, and their survival in diluted media is dependent upon such degree of tolerance of lowered haemolymph concentration as the cells may possess. Usually this is limited and few crustacean species lacking regulatory capacity tolerate extended exposure to water of less than about 70% of the salinity of seawater. (2) All truly estuarine species maintain the haemolymph concentration at a level higher than that of the medium over at least part of the salinity range tolerated though many hypertonic regula-

Fig. 1. Examples of the kinds of relations maintained between blood and medium in crustaceans (see text for details). (1) Osmoconformers would have blood concentrations following the isosmotic line to the limit of their tolerance, e.g. *Maía*. (2) Hypertonic–isotonic regulation, *Gammarus duebeni*. (3) Hypertonic regulation only, *Paragnathia formica*. (After Appelbee, 1975.) (4) Hypotonic–hypertonic regulation, *Sphaeroma rugicauda*. (After Harris, 1967.)

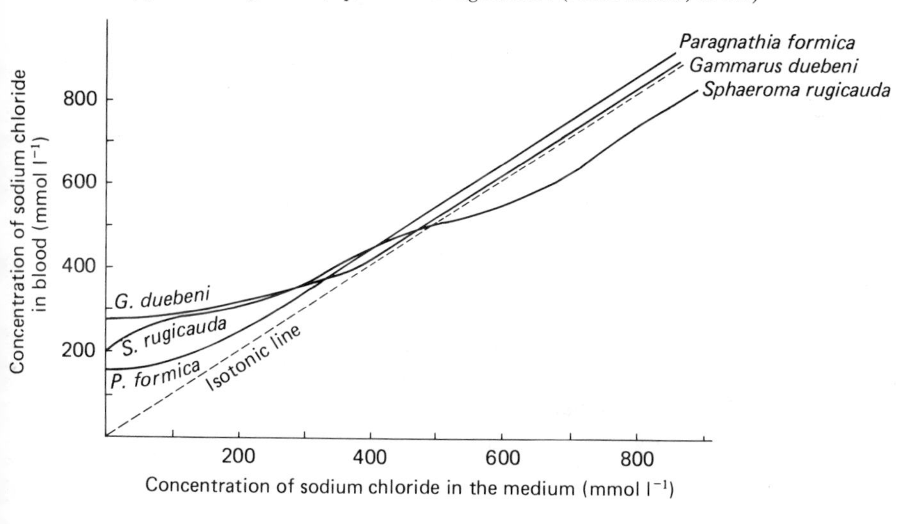

tors have body fluids close to isotonicity when in seawater or water of higher salinity. The range of dilution over which maintenance of hypertonicity is possible varies with species, as does the degree of hypertonicity attained. Freshwater crustaceans form a sub-set of the hypertonic regulators, maintaining blood concentrations at levels substantially above the concentrations of their habitat. They tolerate some rise in external concentration but few species (the lake race of the isopod *Mesidotea entomon* is an exception (Lockwood, Croghan & Sutcliffe, 1976)), survive if the concentration of the medium is raised to a level much above the normal blood concentration. (3) An interesting variant on the hypertonic regulator theme is provided by the ectoparasitic praniza larva of the isopod *Paragnathia*. This lives on the gills of estuarine fish and consequently is liable to experience suddenly raised salinities should the host move seawards. The praniza larvae are small (*c.* 5 mg) and have a high permeability to water. Any increase in external concentration above the level of their body fluids would thus potentially lead to risk of substantial fluid loss by osmosis. The fact that the body fluids are normally regulated at a level some 300 mOsmol above the environmental level provides some buffer against water loss on exposure to more concentrated environments (Appelbee, 1975). (4) The final group, including *Artemia* and a number of isopods, decapods and mysids are hypotonic–hypertonic regulators, able to maintain the body fluids more concentrated than dilute media and less concentrated than strongly saline media.

Regulation of the body fluids at a concentration above that of the medium necessitates the expenditure of energy to remove water taken in by osmosis and to replace inorganic ions lost by diffusion and in the urine. Reduction of this energy expenditure can be achieved either by limiting the osmotic gradient maintained between blood and medium or by restricting the permeability of the body surface so as to decrease the rate of ion and water turnover. Total restriction of surface permeability is impractical because of the requirement to maintain a diffusion pathway for respiratory gas exchange, but in the larger crustaceans the area over which gas exchange occurs is essentially limited to specialised structures (gills or respiratory plaques). Such regions are generally also the part of the body surface (excluding the gut) most permeable to water and ions. For example, the calculated permeability of the general body surface of the crab *Uca pugilator* is 7.5 $\mu m^3\ \mu m^{-2}\ s^{-1}$ by comparison with 83.0 $\mu m^3\ \mu m^{-2}\ s^{-1}$ for the gills (Hannan & Evans, 1973).

The respiratory areas are considered to be the principal site of ion regulation and osmotic water movements, though the evidence for this is direct only in the case of the decapods. Ion regulation in the Crustacea has been extensively reviewed in recent years (Gilles, 1975; Lockwood, 1976, 1977; Kirschner, 1979). This paper will therefore concentrate on aspects relating to the permeability to water and the role which ion-induced water transport may play in volume regulation, intake of water for urine production under isosmotic conditions and expansion at moult. The gills are seen as playing an important role in these functions.

Identification of permeable regions of the body surface

The respiratory and ionoregulatory regions in crustaceans have a wide spectrum of anatomical origins ranging, for example, from the general body surface in some of the smaller forms through specialised plaques on parts of the body surface (mysids and syncarids), and parts of the pleopods and lower telson surface (isopods) to podobranch, arthrobranch and pleurobranch gills of the thoracic appendages (decapods). A classical method of establishing the areas of the body surface sufficiently permeable to permit the ready passage of inorganic ions has been silver staining, and though no reliance can be placed on this technique as a specific means of delimiting regions capable of ion transport, there are cases where good correlations exist. Thus the silver-staining plaques on the dorsal thorax of *Allanaspis* are underlain by hypodermal cells that are morphologically similar to the transporting cells seen in the gills of other crustaceans (Lake, Swain & Ong, 1974). Isolated gill studies (Koch, Evans & Schicks, 1954; Bielawski, 1964; Croghan, Curra & Lockwood, 1965) conclusively demonstrate the ability of decapod gills to transport ions against electrochemical gradients. Ion transport capacity is usually attributed to specialised hypodermal cells on the gills which are characterised by substantial infoldings of the basal membrane and, at least when euryhaline species are in dilute media, by the presence of lamellae at the apical surface (Copeland & Fitzjarrel, 1968; Lockwood, Inman & Courtenay, 1973; Lake *et al.*, 1974).

It has been recognised since the classical studies of Nagel (1934) that the permeability of crustaceans can be correlated to the environment that they inhabit. His experiments indicated that iodine penetrates more rapidly across the body surface of stenohaline marine species than across that of euryhaline estuarine forms. The latter in turn were more permeable than freshwater forms. Gross (1957) confirmed this general relation with regard to sodium movement across the cuticle of various decapods. Similar considerations apply to permeability to water as is indicated by comparison of the water flux of related marine, estuarine and freshwater forms in relation to the osmotic gradient that they maintain. Such data indicate that the permeability of the body surface to water of species inhabiting dilute media is in general lower than that of their more marine relatives (Lockwood, 1977). Kirschner (1979) and Rudy (1967), extending the study of water permeability to water fluxes across the body surface using heavy water ([^{2}H]water), observed that the stenohaline marine crab *Macropipus depurator* had a markedly higher permeability constant than the euryhaline *Carcinus*, which in turn displayed a larger flux than the crayfish. As would be expected, a large part of such water exchange occurs across the gills; some 86–90% of the observed water flux in the crabs *Rhithropanopeus* (Capen, 1972) and *Uca pugilator* (Hannan & Evans, 1973). Gray (1957) discusses the gill area of crabs in relation to the environment they inhabit and Herreid (1969) has examined tegumental permeability in relation to adaptation to a terrestrial mode of life.

A lively issue over the last fifteen years has been the question of whether or not individual animals can actually vary their water permeability in response to a change in the external salinity or other factors. The arguments depend on whether or not flux measurements accurately represent the hydraulic permeability of the body surface. Before considering the potential problems associated with interpretation of water fluxes, however, it is perhaps appropriate to examine the factors which have been observed to affect flux constants.

Smith (1967) and Smith & Rudy (1972) demonstrated that the permeability constant for water flux in the estuarine crabs *Rhithropanopeus harrisi* and *Hemigrapsus nudus* declines with decreasing salinity, though these authors expressed a proper caution in interpreting this effect in terms of an actual change in permeability referring only to 'apparent permeability'. The same terminology is adopted here.

Since Smith's pioneer study, a number of other factors have been tested with regard to their influence on apparent permeability to water. In addition to salinity these include osmotic concentrations, substitution or addition of ion species, temperature, body size, handling stress, potential difference changes, moult and neural extracts.

Factors influencing apparent water permeability

Salinity

A wide range of euryhaline crustaceans have been observed to show reductions in apparent permeability on dilution (the prawn *Palaemonetes pugio* (Roesijadi, Anderson, Petrocelli & Giam, 1976), the crabs *Cancer irroratus* and *Callinectes sapidus* (Cantelmo, 1977) and the amphipod *Gammarus duebeni* (Lockwood *et al.,* 1973)). In addition, some hypotonic regulators such as *Artemia salina* (Stewart, 1974) and *Palaemonetes pugio* (Roesijadi *et al.,* 1976) show decreased water flux exchange constants with increasing salinity gradient when they are hypotonic. A similar effect is observed in euryhaline teleosts which also tend to have lower exchange constants in raised salinities (Potts, Foster, Rudy & Parry Howells, 1967; Evans, 1969; Motais, Isaia, Rankin & Maetz, 1969).

Not all crustaceans, however, show clear-cut variations in apparent permeability when exposed to a change in the salinity of their environment. Included in this category are *Uca pugilator, U. minax, U. rapax, Pennaeus duorarum* (Hannan & Evans, 1973) and *Palaemonetes varians* (Rudy, 1967).

One of the larger reported changes in apparent permeability with salinity of individuals that are fully acclimated to their medium is that shown by *Artemia salina*, where the half-time ($T_{\frac{1}{2}}$) for exchange of tritiated water varies by a factor of just over three in the salinity range 38% seawater to 566% seawater (Stewart, 1974). In this animal the water flux constant seems to reflect the osmotic gradient between blood and medium. By contrast, the effect in *Gammarus duebeni* appears to be non-linearly related to the osmotic gradient. For animals of this species acclimated to a range of salinities the largest part of the change in the rate constant for water flux occurs in the range 50–70% seawater (Fig. 2) (Lockwood *et*

Fig. 2. (*a*) The relation between blood concentration and medium concentration of sodium chloride in acclimated *G. duebeni*. (*b*) Half-times for [^{3}H]water exchange in *G. duebeni* acclimated to various salinities in the same range.

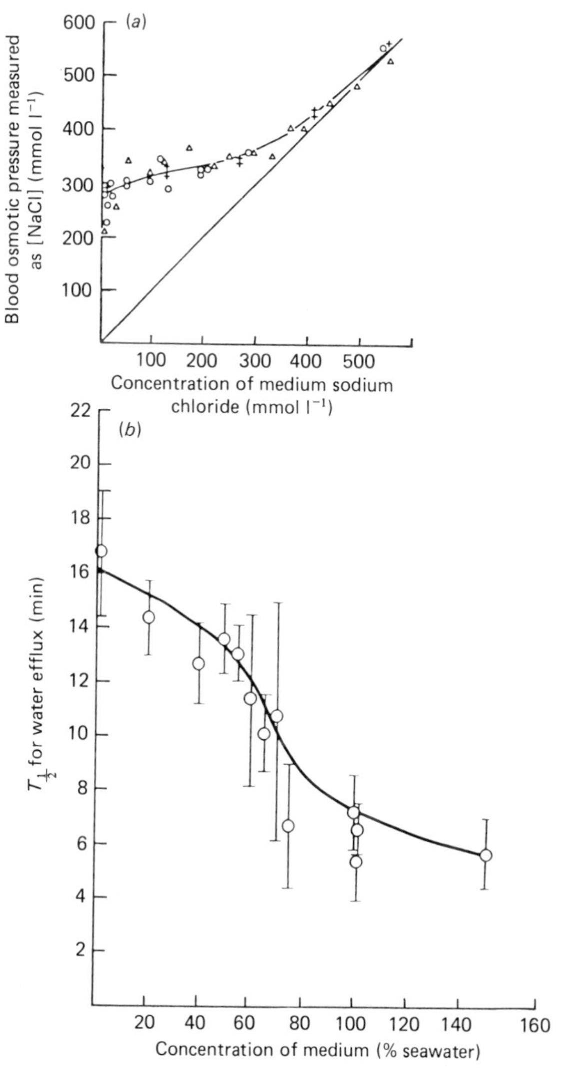

al., 1973; Bolt, Dawson, Inman & Lockwood, 1980) though the degree of hypertonicity of the body fluids is small in this part of the salinity range. A similar non-linear effect in respect of variation in apparent permeability has been observed for the isopod *Cyathura carinata* (Hussain, 1973).

Salinity 'shock' effects in Gammarus duebeni. Sudden change in the salinity of the medium which causes the body fluid of *Gammarus* to become either hypertonic or hypotonic results in a decrease in the apparent permeability level observed in individuals acclimated to media with which they are isosmotic (Lockwood *et al.*, 1973). The time scale of the change from one exchange constant to the new state is short, being less than about one minute in each case (Fig. 3).

No such salinity shock effects have been seen in crabs. Isolated, perfused gills of *Callinectes* and *Cancer* acclimated to 100% seawater have the same apparent permeability subsequent to the tests whether the test medium is 100% seawater or 10% seawater. Likewise, gills of crabs initially acclimated to 40% seawater have similar flux constants in 40% and 4% seawater (Cantelmo, 1977).

Longer-term changes in apparent permeability. Gradual changes in apparent permeability can be classified under two headings; (a) those occurring in the initial period of acclimation to a new medium whilst blood concentration is

Fig. 3. The effect of a sudden increase or decrease in medium concentration on water flux in *Gammarus duebeni*. Open circles, half-time for [^{3}H]water exchange in animals initially acclimated to 100% seawater and transferred to 2% seawater; closed circles, half-time for [^{3}H]water exchange in individuals acclimated to 2% seawater and transferred to 100% seawater after the initial measurement.

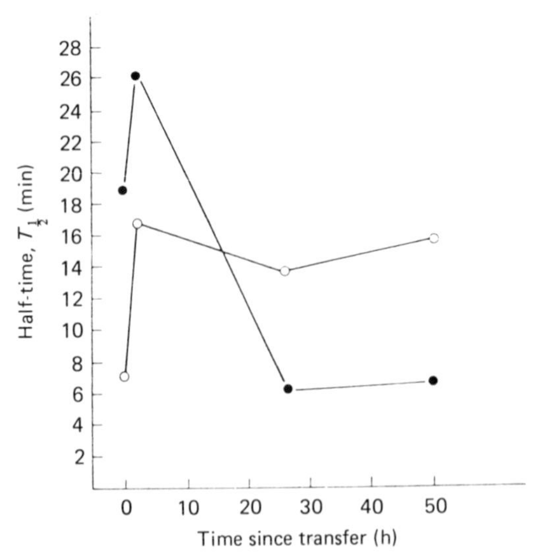

changing and (b) long-term effects, possibly associated with modification of membrane lipid composition.

(a) Transference of crabs from one medium to another, although not producing immediate shifts in water flux constants, does result in changes during the period in which the blood concentration is adjusting towards a new steady state. Transference of *Rhithropanopeus* from 75% to 10% seawater, or in reverse from 10% to 75% seawater, epitomises this effect. Variation in apparent permeability more or less parallels the adjustment in blood concentration (Capen, 1972) (Fig. 4). Similarly, *Callinectes* and *Cancer irroratus*, though not displaying immediate changes in water flux after transfer from one medium to another, do show differences after acclimation (Cantelmo, 1977).

In complete contrast to the *Rhithropanopeus* effect is the response of *Gammarus duebeni* when exposed to a salinity higher than the blood concentration (Fig. 5). Initially, on transference to hypertonic saline, the apparent permeability suddenly decreases. This new level is maintained as the blood concentration climbs towards isotonicity but at about the time isotonicity is reached there is a sudden increase in the flux constant.

(b) The influence of a long period of acclimation to a dilute medium on water fluxes appears to have been examined only in *Gammarus duebeni*.

Fig. 4. The relation between the rate of water exchange and blood chloride concentration in the crab *Rhithropanopeus harrisi* after transfer from 10% to 75% seawater. (From Capen, 1972.)

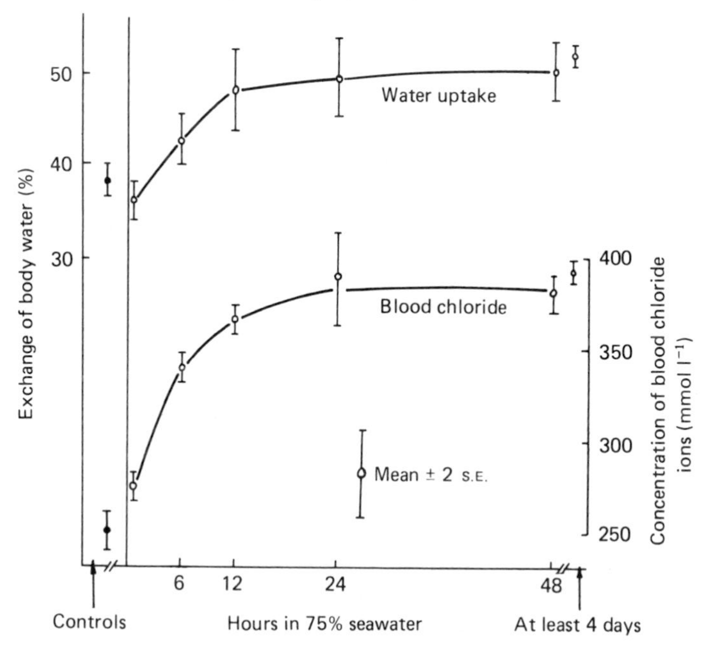

G. duebeni which have been maintained in 2% seawater for an extended period of time (two months) are less permeable than freshly-caught controls following three days of acclimation to 2% seawater (Table 1). Also, individuals initially acclimated for six weeks to 100% seawater tend to have shorter half-times for exchange than controls in 2% seawater for six weeks (Table 1). Since both the flux at a given concentration and the response to a new medium have changed following acclimation the most obvious interpretation of these differences is that long-term acclimation to particular salinities may result in structural modifications which influence permeability. Potentially, such effects could involve either modification of membrane structure in the gills or, by analogy with the hydroid *Cordylophora* (Kinne, 1958), a change in cell dimensions.

Comparative responses to salinity change. Animals living in bodies of water which are liable to experience rapid and extensive salinity changes, (e.g. the pools and creeks of salt marshes and rock pools at the levels reached by spring

Fig. 5. (*a*) Blood sodium concentration at various times after transfer of *Gammarus duebeni* from 2% to 100% seawater. Full line, blood sodium concentration, [Na] ± s.e.m.; interrupted line, medium sodium concentration, [Na]. (*b*) Half-time for water exchange after the transfer. Full line, mean half-time, $T_{1/2}$; crosses, individual half-times for apparent permeabilities. Note the sudden change in half-time for water exchange occurring after about 15 h.

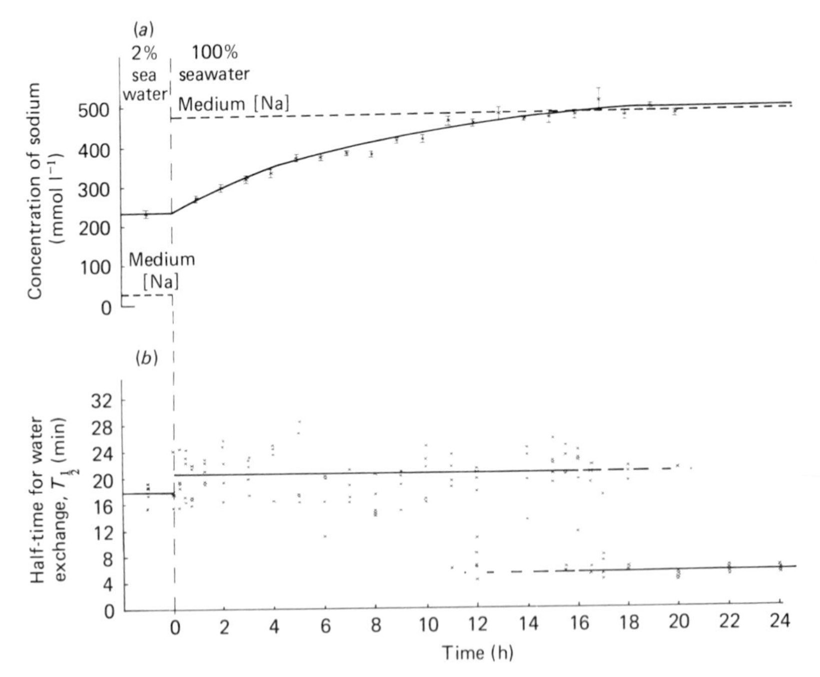

tides) are likely to be amongst the forms which would benefit most from the capacity to effect a rapid permeability change. Such a response would permit time for physiological reactions appropriate to the new conditions to be mobilised by limiting water loss or uptake in the short term. Comparison of similar-sized forms from different natural habitats tends to confirm that changes in the $T_{\frac{1}{2}}$ for [^{3}H]water exchange are more pronounced in a species from such a habitat, e.g. *G. duebeni*, than in related estuarine and marine crustaceans. Thus when *G. duebeni*, *Chaetogammarus* (*Marinogammarus*) *marinus*, *Gammarus locusta* and *Corophium volutator* are exposed to a salinity cycle varying between 3% and 97% seawater with a 12h 25 min periodicity, the changes in apparent permeability shown by *G. duebeni* are substantially greater than those displayed by the estuarine gammarids or by the burrowing *Corophium* (Fig. 6). The nature and direction of the changes in apparent permeability shown by *Gammarus* would, if the $T_{\frac{1}{2}}$ variation reflects genuine permeability change, be appropriate in direction to limit water transference across the body surface when individuals are either strongly hypotonic or strongly hypertonic to the medium.

Table 1. *Half-time for water exchange in long-term and short-term acclimated* Gammarus duebeni. Values are $T_{\frac{1}{2}}$ (min) for [^{3}H] water exchange ± s.e.m.

	Test medium	
	2% seawater	100% seawater
Freshly caught animals after 3 days in 2% seawater	17.3+0.7 (12)	6.2±0.3 (12)
Animals after 66 days in 2% seawater	23.4±0.8 (12)	7.3±0.4 (12)
Controls initially acclimated to 2% seawater for 46 days	19.1±0.7	—
	P = <0.05	—
Animals acclimated to 100% seawater for 46 days then transferred to 2% seawater for 3 days	16.5±0.9	—

Osmotic concentration

Substitution of non-electrolytes for all or part of the ionic components of the medium results in a change in the water flux constant. *Gammarus duebeni* acclimated to 100% seawater and transferred to a mannitol solution that is isosmotic with seawater display a decrease in the water flux (Lockwood & Inman, 1973*a*). The half-time for [^{3}H]water exchange increases from the seawater level of about five to six minutes to nearer the level observed in 2% seawater. Similarly exposure of *Rhithropanopeus* to 10% seawater made isosmotic with 100% seawater using sucrose results in a decline in the apparent water permeability (Capen, 1972). It may be concluded that the osmotic pressure *per se* is not the determinant of the level of apparent permeability.

Ion substitution

Substitution of either chloride or sodium alone in the medium has no significant effect on the water exchange rate in *Rhithropanopeus* (Capen, 1972).

Modification of the divalent ion level in the medium produces variable effects. A decrease in apparent permeability of *Artemia salina* is associated with an increase in the magnesium concentration of the medium but, by contrast, lowering the level of calcium has little effect (Stewart, 1974). Similarly calcium has little effect on the water fluxes of *Chirocephalus diaphanus*, the exchange being identical in this freshwater branchiopod whether the medium is 1 mmol l^{-1} NaCl or 1 mmol l^{-1} NaCl + 2 mmol l^{-1} $CaCl_2$ (Stewart, 1974).

Fig. 6. The half-time for water ([^{3}H]water) exchange of four amphipod species exposed to a cycling salinity regime. Interrupted line, medium sodium concentration.

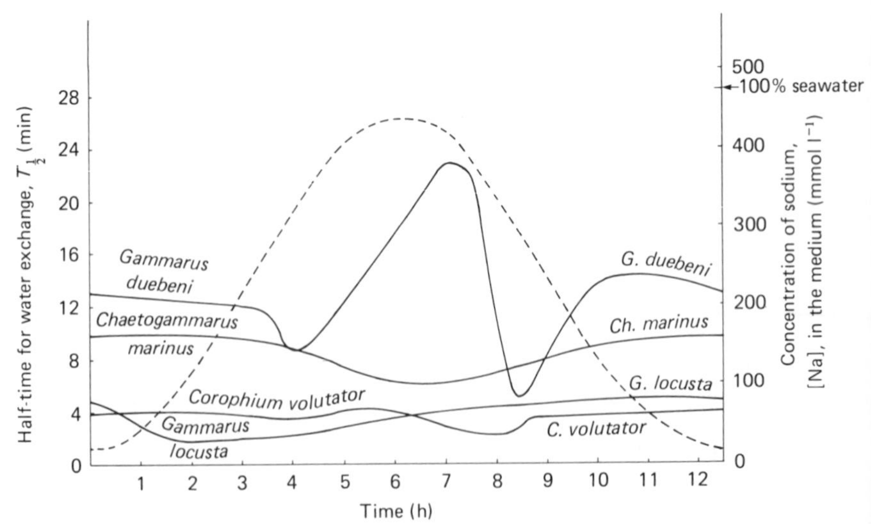

The water fluxes of the prawn *Palaemonetes varians* are, however, substantially affected by depletion of calcium in the medium (Table 2).

Temperature

The Q_{10} for tritiated water ([^{3}H] water) and deuterated water ([^{2}H] water) influx in the crab *Hemigrapsus nudus* is of the order of 1.6 in the temperature range 10–20°C (Smith & Rudy, 1972) and values for *Palaemonetes varians* and *Artemia salina* fall in the range 1.56–2.76 (Stewart, 1974; Tin Tun, 1975). There is some suggestion that Q_{10} declines with salinity in the case of *Uca pugilator* (Hannan & Evans, 1973) and with temperature in *Cyathura carinata* (Hussain, 1973).

Body size

The relation between body size and water flux can be expressed as $M + aW^X$ where M is the influx, a is the intercept on the y-axis, W is the body weight and X is the slope of the line. Values of X recalculated by Hannan & Evans (1973) from the data of Smith (1970) include 0.81 for *Rhithropanopeus* and 0.73 for *Carcinus*.

Handling and other stress factors

Water influx does not change in *Uca pugilator* as a result of handling stress or autotomisation of a walking leg (Hannan & Evans, 1973).

Moult

Water flux approximately doubles at ecdysis in both *Gammarus duebeni* and *Idotea linearis*. In the former species the factor by which flux increases is apparently independent of the medium in which moult occurs. The flux constant

Table 2. *The influence of calcium on water flux in* Palaemonetes varians *at 14 °C (Tin Tun, 1975)*

	$T_{\frac{1}{2}}$ (min)	(n)
100% Seawater (Ca^{2+}-free)	15.3 ± 1.1	(7)
100% Seawater (0.1 mmol l^{-1} Ca^{2+})	18.2 ± 2.3	(6)
100% Seawater (1.0 mmol l^{-1} Ca^{2+})	20.0 ± 2.1	(7)
100% Seawater (10.0 mmol l^{-1} Ca^{2+})	22.0 ± 1.4	(7)
2% Seawater (Ca^{+}-free)	14.5 ± 1.1	(6)
2% Seawater (0.1 mmol l^{-1} Ca^{2+})	19.5 ± 1.8	(8)
2% Seawater (1.0 mmol l^{-1} Ca^{2+})	20.3 ± 1.8	(6)
2% Seawater (10.0 mmol l^{-1} Ca^{2+})	21.4 ± 2.0	(8)

returns to the normal level after approximately four days (Lockwood & Inman, 1973*b*).

Neural extracts

Limited studies have been undertaken to investigate possible hormonal control of water permeability. Neither loss of eyestalks nor injection of extracts of the sinus gland or pericardial glands (Berlind & Kamemoto, 1977) affects the capacity of *Carcinus maenas* to reduce apparent permeability on dilution of the medium. The same authors find, however, that aqueous extracts of thoracic glands cause a decrease in [^{3}H]water influx when perfused through isolated gills of *Carcinus*. However, as Berlind & Kamemoto note, the concentration of extract in the perfusate was high and there is no certainty that the response is to a hormone *per se*. Thoracic ganglion extracts have been reported to increase the permeability of the foregut of *Gecarcinus* to water (Mantel, 1968).

It may be safely concluded that there are substances in the CNS which are capable of increasing the permeability of crustacean epithelia. However, as no successful attempts appear to have been made as yet to demonstrate the presence of such an agent in the haemolymph, it remains uncertain whether these substances are (a) the actual effector agents of permeability change in intact crustaceans and (assuming they are), (b) delivered via the haemolymph or by some more localised neurosecretory mechanisms.

The various findings that water flux constants obtained using tritiated or deuterated water can be caused to vary raises the pertinent question as to whether or not changes in the constant represent genuine variation in hydraulic permeability.

Do differences in water flux represent real differences in hydraulic permeability or artefacts of measurement?

There are essentially three possible interpretations of variations in water flux.

(a) The differences are artefacts of measurement not associated with change in the hydraulic permeability of the epithelium, i.e. they do not reflect changes in the epithelium which affect the impedance to water movement.

(b) Flux variations are caused by variation in the surface area over which water movement can occur due to circulatory re-circuiting within the gill. No change in the hydraulic permeability of the unaffected exchange surface is postulated.

(c) Real variations in the hydraulic permeability occur and are reflected in changes in $T_{\frac{1}{2}}$ for water exchange.

Should (a) be the explanation of flux differences the effects are of no biological significance and can be ignored in any assessment of biological response to salinity. By contrast both (b) and (c) could represent biological responses which, in view of the direction of the observed changes, could confer energetic advantages. (c) alone involves changes at the intracellular level.

Possible artefacts of measurement (a)

As pointed out by Dainty & House (1966), variation in the thickness of unstirred layers on either side of a membrane will be likely to result in apparent differences in the rate of transfer of isotopically labelled molecules [^{3}H] water or [^{2}H] water between the bulk phase of the two solutions. Similarly the passage of water through narrow pores in a membrane might be expected to result in distortions of the theoretical diffusional exchange from one side of a membrane to the other (Koefoed-Johnsen & Ussing, 1953).

The ratio of the osmotic permeability (p_{os}, as determined from the net transfer of water) and the diffusional permeability (p_d determined from the flux of [^{2}H] water or [^{3}H] water) can differ markedly. In the case of membranes such as the frog skin, which lack overt irrigation systems on both sides of the epithelium, permeability measurements based on flux determinations alone can substantially underestimate the net passage of water down an osmotic gradient. Thus the p_{os}/p_d ratio for amphibian skin can exceed ten (*Bufo regularis*, 27.2; *Xenopus laevis*, 9.3; *Rana esculenta*, 10.1) (Maetz, 1968). By contrast irrigated tissues such as gills generally have lower p_{os}/p_d ratios whether they are of vertebrate or invertebrate origin. (The teleost fish *Anguilla* and *Platichthys* have respective ratios of 3.16 and 2.59 when in freshwater and 1.05 and 0.83 when in seawater (Motais *et al.*, 1969) whilst for *G. duebeni* the corresponding values are 1.16 in 2% seawater and 1.97 in 100% seawater.) Similarly the crabs *Libinia emarginata*, *Pugettia producta* and *Carcinus maenas* give p_{os}/p_d ratios ranging between 1.0 and 2.5 depending on which figures are used for urine flow (Cornell, 1979*b*). Possibly the higher value should be diminished somewhat as there is now evidence that the use of tracer markers to determine the urine flow in crabs can in some cases overestimate urine flow where there is reabsorption of water but not of the marker in the excretory system (Riegel *et al.*, 1974; Zanders, 1980).

Presumably the lower p_{os}/p_d ratio observed in gills can be ascribed to the irrigation of the outside by the respiratory currents and of the inside by blood, so that unstirred layers are diminished.

In some species the decrease in apparent permeability commonly observed on transfer of an individual from a high salinity to a lower one may be attributed to a change in heart rate, and hence irrigation of the inner surface of the gill epithelium. Thus Cornell (1979*a*) reports that transfer of the crab *Libinia* from 100% to 80% seawater is followed by a reduction in heart rate which seems to correlate with the observed change in water flux (Cornell, 1979*b*).

However, in forms more euryhaline than *Libinia* it seems less likely that the changes observed in water flux as the media is varied can be attributed to differences in heart rate or respiratory current generation. *Gammarus duebeni* shows only small changes in heart rate when transferred successively from 100% seawater to freshwater and the differences that do occur are such that heart rate is faster in the dilute medium (Bolt *et al.*, 1980). Similarly, the heart rate of *Carcinus maenas* increases on acclimation to dilute media (Hume & Berlind, 1976).

The general conclusion to be drawn from studies of heart rate and respiratory current generation is that it is unlikely that unstirred layers will increase in thickness in dilute media in euryhaline species as a result of decreased flow over either side of the gill surface. Differences in diffusional permeability as the medium is changed cannot therefore readily be ascribed to this course.

A further factor possibly contributing to p_{os}/p_d differences is isotopic disparity between the diffusion rates of [^{3}H]water, [^{2}H]water and [^{1}H]water. Smith (1976) suggests that the true diffusion value for water (H_2O) will be underestimated by some 5–10% by [^{3}H]water flux measurements.

Circulatory regulation in the gills (*b*)

Flux changes would be expected to occur even in the absence of actual permeability change if the surface area available for exchange is altered. Support for excluding circulatory restriction in the gill as a cause of decreased apparent permeability to water comes from studies on *Rhithropanopeus* where oxygen uptake increases in dilute media despite the concomitant decrease in flux constant (Capen, 1972).

In view of the lack of positive evidence that passive factors could offer an explanation of changes in apparent permeability as large as those observed, there is some justification in examining the possible means by which an actual change in water permeability might be achieved, though without prejudice to the issue as to whether such factors are indeed involved in producing the observed flux effects.

Real changes in hydraulic permeability (*c*)

Prior to consideration of the mechanisms which could potentially be involved in restriction of water flow we must summarise current views of the routes by which water may traverse epithelia. Two fundamental pathways are possible: (1) via the cells and (2) via the intercellular channels.

(1) *Passage via the cell.* This necessitates the traversing of the plasma membrane twice. Water could theoretically pass (i) via a solubility–diffusion mechanism through the lipid bilayer of the plasma membrane, (ii) by pores with hydrophilic cores traversing the membrane or (iii) by pinocytosis.

Passage of water by a solubility–diffusion mechanism through artificial phospholipid bilayers will give rise to p_{os}/p_d ratios of *c.* one after correction for unstirred layers (Holz & Finkelstein, 1970; Finkelstein, 1976*a,b*). By contrast, movement through pores would be expected to generate p_{os}/p_d ratios in excess of unity and, since bulk flow (and hence osmotic permeability, p_{os}) increases as the square of the pore area, whereas diffusion (and hence p_d) increases in direct proportion to the surface area (Oschman, Wall & Gupta, 1974), any enlargement of pores will tend to enhance the p_{os}/p_d ratio. Pores through the basal, apical and lateral membranes could occur, but if they do there is as yet no positive evidence for their role in relation to water movements.

(2) *Passage via intercellular channels.* Vertebrate studies have suggested the possibility that in some tissues, e.g. the gall bladder (Hill, 1978), and corneal membrane (Fischbarg, 1978), fluid may pass via the intercellular channels. Again there is unfortunately no information for crustaceans as to whether this route is significant in the gills though the relative impermeability of gill tissues and the pavement-like nature of the epithelial transporting cells perhaps militates against the concept.

As in so many fields, crustacean osmoregulatory physiology lags behind vertebrate studies and nowhere is this more lamentably obvious than in the problems relating to permeability and water movement. Even to the most casual reader of the literature in the field it must be apparent that, while most of those who have studied the water fluxes of the more euryhaline members of the group would like to feel that short-term changes in apparent permeability represent a genuine ability to modify hydraulic permeability, tangible evidence to that effect has proved elusive. Certainly, seen in the context of animals such as *Gammarus duebeni* which are potentially liable to suffer rather rapid and extreme changes in the concentration of the small bodies of water in which they live, it would seem, from a teleological viewpoint, to be desirable for them to be able to restrict transepithelial water movements when the body fluids are markedly hypotonic or hypertonic to the medium. Evidence from two sources now suggests that changes in the apparent permeability with salinity obtained by [^{3}H] water fluxes may reflect, albeit perhaps not mirror, corresponding changes in hydraulic permeability. This evidence is based upon (a) comparison of urine flow ([^{51}Cr] EDTA clearance studies) of individuals in cycling salinity regimes where clearance, although broadly following the osmotic gradient, is not proportional to it (S.R.L. Bolt, unpublished observations) and (b) the observation that the sodium concentration of the blood declines markedly less rapidly (about three times) than would be expected from the gradient in the first hour after transference from 100% seawater to 2% seawater. Even allowing for water movement to the cells and increased ion uptake at the body surface, the discrepancy is too large to be explained unless water entry per unit osmotic gradient is less than the level pertaining in individuals acclimated to seawater.

Evidence is thus accruing which may eventually lead to the acceptance of the possibility that at least some crustaceans may be capable of varying hydraulic permeability as a response to environmental conditions. More detailed studies of the precise relation between urine flow and [^{3}H] water flux measurements are still needed, however, before the pejorative adjective can finally be dropped from the term 'apparent permeability change'.

The kind permission of Dr R. L. Capen and of the Journal of Experimental Zoology for the use of Fig. 4 is gratefully acknowledged.

This review was written whilst M.E.D. was employed on a Natural Environment Research Council (NERC) grant to A.P.M.L. and S.R.L.B. was a NERC-supported student.

References

Appelbee, J. (1975). An investigation into some aspects of the osmoregulation of *Paragnathia formica* (Hesse). M.Sc. Dissertation, University of Southampton.

Berlind, A. & Kamemoto, F. I. (1977). Rapid water permeability changes in eyestalkless euryhaline crabs and in isolated perfused gills. *Comparative Biochemistry and Physiology,* **58A,** 383–5.

Bielawski, H. (1964). Chloride transport and water intake into isolated gills of crayfish. *Comparative Biochemistry and Physiology,* **13,** 423–32.

Bolt, S. R. L., Dawson, M. E., Inman, C. B. E. & Lockwood, A. P. M. (1980). Variation of apparent permeability to water and sodium transport in *Gammarus duebeni* exposed to fluctuating salinities. *Comparative Biochemistry and Physiology,* **67B,** 465–73.

Cantelmo, A. C. (1977). Water permeability in isolated tissues from decapod crustaceans. I. Effect of osmotic conditions. *Comparative Biochemistry and Physiology,* **58A,** 343–8.

Capen, R. L. (1972). Studies of water uptake in the euryhaline crab, *Rhithropanopeus harrisi. Journal of Experimental Zoology,* **182,** 307–20.

Copeland, D. E. & Fitzjarrel, A. T. (1968). The salt absorbing cells in the gills of the blue crab *Callinectes sapidus* (Rathbun) with notes on modified mitochondria. *Zeitschrift für Zellforschung und Mikroskopische Anatomie,* **92,** 1–22.

Cornell, J. C. (1979*a*). Salt and water balance in two marine spider crabs, *Libinia emarginata* and *Pugettia producta.* I. Urine production and magnesium regulation. *Biological Bulletin of the Marine Biology Laboratories, Woods Hole,* **157,** 221–33.

Cornell, J. C. (1979*b*). Salt and water balance in two marine spider crabs, *Libinia emarginata* and *Pugettia producta.* II. Apparent water permeability. *Biological Bulletin of the Marine Biology Laboratories, Woods Hole,* **157,** 422–33.

Croghan, P. C., Curra, R. A. & Lockwood, A. P. M. (1965). The electrical potential difference across the epithelium of isolated gills of the crayfish, *Austropotamobius pallipes* (Lereboullet). *Journal of Experimental Biology,* **42,** 463–74.

Dainty, J. & House, C. R. (1966). 'Unstirred layers' in frog skin. *Journal of Physiology,* **182,** 66–78.

Evans, D. H. (1969). Studies on the permeability to water of selected marine, freshwater and euryhaline teleosts. *Journal of Experimental Biology,* **50,** 689–703.

Finkelstein, A. (1976*a*). Water and non-electrolyte permeability of lipid bilayer membranes. *Journal of General Physiology,* **68,** 127–35.

Finkelstein, A. (1976*b*). Nature of the water permeability increase induced by antidiuretic hormone (ADH) in toad bladder and related tissues. *Journal of General Physiology,* **68,** 137–43.

Fischbarg, J. (1978). Fluid transport by corneal epithelium. In *Comparative Physiology: Water, Ions and Fluid Mechanics,* ed. K. Schmidt-Nielsen, L. Bolis & S. H. P. Maddrell. Cambridge University Press.

Gilles, R. (1975). Mechanisms of ion and osmoregulation. In *Marine Ecology,* ed. O. Kinne. London: John Wiley and Sons.

Gray, I. E. (1957). A comparative study of gill areas of crabs. *Biological Bulletin of the Marine Biology Laboratories, Woods Hole,* **112,** 34–42.

Gross, W. J. (1957). An analysis of response to osmotic stress in selected decapod crustacea. *Biological Bulletin of the Marine Biology Laboratories, Woods Hole,* **112,** 43–62.

Hannan, J. & Evans, D. H. (1973). Water permeability in some euryhaline decapods and *Limulus polyphemus. Comparative Biochemistry and Physiology,* **44A,** 1199–213.

Harris, R. R. (1967). Aspects of ionic and osmotic regulation in two species of *Sphaeroma* (Isopoda). Ph.D. Thesis, University of Southampton.

Herreid, C. F. (1969). Integumental permeability of crabs and adaptation to land. *Comparative Biochemistry,* **29,** 423–9.

Hill, A. E. (1978). Fluid transport across *Necturus* gall bladder epithelium. In *Comparative Physiology: Water, Ions, and Fluid Mechanics,* ed. K. Schmidt-Nielsen, L. Bolis & S. H. P. Maddrell. pp. 360. Cambridge University Press.

Holz, R. & Finkelstein, A. (1970). The water and non-electrolyte permeability induced in thin membranes by the polyene antibiotics nystatin and amphotericin. *Journal of General Physiology,* **56,** 125–45.

Hume, R. I. & Berlind, A. (1976). Heart and scaphognathite rate changes in a euryhaline crab, *Carcinus maenas* exposed to a dilute environmental medium. *Biological Bulletin of the Marine Biology Laboratories, Woods Hole,* **150,** 241–54.

Hussain, N. A. (1973). The influence of environmental salinity and temperature on the water fluxes of the isopod crustacean *Cyathura carinata* (Kryer). M.Sc. Dissertation, University of Southampton.

Kinne, O. (1958). Adaptation to salinity variations – some facts and problems. In *Physiological Adaptations,* ed. C. L. Prosser. Washington: American Physiological Society.

Kirschner, L. B. (1979). Control mechanisms in crustaceans and fishes. In *Mechanisms of Osmoregulation in Animals,* ed. R. Gilles. Chichester: John Wiley and Sons Limited.

Koch, H. J., Evans, J. & Schicks, E. (1954). The active absorption of ions by the isolated gills of the crab *Eriocheir sinensis* (M.EdW) *Mededelingen van de K. Vlaamsche Academie voor Wetenschapen, Letteren en Schoone Kunsten van België. Klasse der Wetenschapen,* **16,** 1–16.

Koefoed-Johnsen, V. & Ussing, H. H. (1953). The contribution of diffusion and flow in the passage of D_2O through living membranes. Effect of hypophyseal hormone on anuran skins. *Acta Physiologica Scandinavia,* **28,** 60–76.

Lake, P. S., Swain, R. & Ong, J. E. (1974). The ultrastructure of the fenestra dorsalis of the syncarid crustaceans *Allanaspides helonomus* and *Allanaspides hickmani*. *Zeitschrift für Zellforschung und Mikroskopische Anatomie,* **147,** 335–57.

Lockwood, A. P. M. (1976). Physiological adaptation to life in estuaries. In *Adaptation to Environment,* ed. R. C. Newell, pp. 315–92. London: Butterworth.

Lockwood, A. P. M. (1977). Transport and regulation in crustacea. In *Transport of Ions and Water in Animals,* ed. B. L. Gupta, R. B. Moreton, J. L. Oschman, & B. J. Wall. New York: Academic Press.

Lockwood, A. P. M., Croghan, P. C. & Sutcliffe, D. W. (1976). In *Perspectives in Experimental Biology,* ed. P. Spencer Davies, vol. 1. Oxford: Pergamon Press.

Lockwood, A. P. M. & Inman, C. B. E. (1973*a*). Water uptake and loss in relation to the salinity of the medium in the amphipod crustacean *Gammarus duebeni*. *Journal of Experimental Biology,* **58,** 149–63.

Lockwood, A. P. M. & Inman, C. B. E. (1973*b*). Changes in the apparent permeability to water at moult in the amphipod *Gammarus duebeni* and the isopod *Idotea linearis*. *Comparative Biochemistry and Physiology,* **44A,** 943–52.

Lockwood, A. P. M., Inman, C. B. E. & Courtenay, T. H. (1973). The influence of environmental salinity on the water fluxes of the amphipod crustacean *Gammarus duebeni*. *Journal of Experimental Biology,* **58,** 137–48.

Maetz, J. (1968). Salt and Water metabolism. In *Perspectives in Endocrinology. Hormones in the lives of lower vertebrates,* ed. E. J. W. Barrington & C. B. Jørgensen, pp. 47–162. New York: Academic Press.

Mantel, L. H. (1968). The foregut of *Gecarcinus lateralis* as an organ of salt and water balance. *American Zoologist,* **8,** 433–42.

Motais, R., Isaia, J., Rankin, J. C. & Maetz, J. (1969). Adaptive changes of the water permeability of the teleostan gill epithelium in relation to external salinity. *Journal of Experimental Biology,* **51,** 529–46.

Nagel, H. (1934). Die Aufgaben der Excretionsorgane und der kiemen bei der Osmoregulation von *Carcinus maenas*. *Zeitschrift für Zellforschung und Mikroskopische Anatomie,* **21,** 468–91.

Oschman, J. L., Wall, B. J. & Gupta, B. L. (1974). Cellular basis of water transport. In *Transport at the Cellular Level,* ed. M. A. Sleigh & D. H. Jennings. *Symposia of the Society for Experimental Biology,* **28.** Cambridge University Press.

Potts, W. T. W., Foster, M. A., Rudy, P. P. & Parry Howells, G. (1967). Sodium and water balance in the cichlid teleost, *Tilapia mossambica*. *Journal of Experimental Biology,* **47,** 461–70.

Riegel, J. A., Lockwood, A. P. M., Norfolk, J. R. W., Bulleid, N. C. & Taylor, P. A. (1974). Urinary bladder volume and the reabsorption of water from the urine of crabs. *Journal of Experimental Biology,* **60,** 167–81.

Roesijadi, G., Anderson, J. W., Petrocelli, S. R. & Giam, C. S. (1976). Osmoregulation of the grass shrimp *Palaemonetes pugio* I. Effects on chloride and osmotic concentrations and chloride- and water-exchange kinetics. *Marine Biology,* **38,** 343–55.

Rudy, P. P. (1967). Water permeability in selected decapod crustacea. *Comparative Biochemistry and Physiology,* **22,** 581–9.

Smith, R. I. (1967). Osmotic regulation and adaptive reduction of water permeability in a brackish-water crab, *Rithropanopeus harrisi* (Brachyura, Xanthidae). *Biological Bulletin of the Marine Biology Laboratories, Woods Hole,* **133,** 643–58.

Smith, R. I. (1970). The apparent water permeability of *Carcinus maenas* (Crustacea, Brachyura, Portunidae) as a function of salinity. *Biological Bulletin of the Marine Biology Laboratories, Woods Hole,* **139,** 351–62.
Smith, R. I. (1976). Apparent water-permeability variation and water exchange in crustaceans and annelids. In *Perspectives in Experimental Biology,* ed. P. Spencer Davies, vol. 1, pp. 17–24. Oxford: Pergamon Press.
Smith, R. I. & Rudy, P. P. (1972). Water exchange in the crab *Hemigrapsus nudus* measured by use of deuterium and tritium oxides as tracers. *Biological Bulletin of the Marine Biology Laboratories, Woods Hole,* **143,** 234–46.
Stewart, A. J. (1974). The influence of environmental salinity, temperature and ionic composition on the water fluxes of *Artemia salina* (L). M.Sc. Dissertation. University of Southampton.
Tin Tun, M. (1975). The influence of environmental calcium and temperature on the water fluxes of *Palaemonetes varians* (Leach). M.Sc. Dissertation, University of Southampton.
Zanders, I. P. (1980). Regulation of blood ions in *Carcinus maenas* (L). *Comparative Biochemistry and Physiology,* **65A,** 97–108.

DAVID H. EVANS

Salt and water exchange across vertebrate gills

The morphological attributes of the vertebrate gill which define its usefulness in gas exchange (i.e. a highly vascularised, thin epithelium with a large surface area) also dictate that it is the site of the majority of any net water and salt movements which may take place if the organism is in an aquatic environment whose osmotic and ionic concentrations differ from those of the body fluids. The evolution of the early vertebrates involved invasion of freshwater environments whose ionic concentrations were lower than those of the marine environments of the protochordates (Fig. 1). The only exception to this pattern is the marine agnathan hagfish which has apparently never entered freshwater (Hardisty, 1979). Thus the early vertebrates were distinctly hyperosmotic (blood more ionically concentrated than the environment) to the surrounding brackish and freshwater very early in their evolution. Since all vertebrates (with the exception of the

Fig. 1. The salinities involved during the early evolution of the vertebrates. See text for details. (Adapted from Lutz, 1975.)

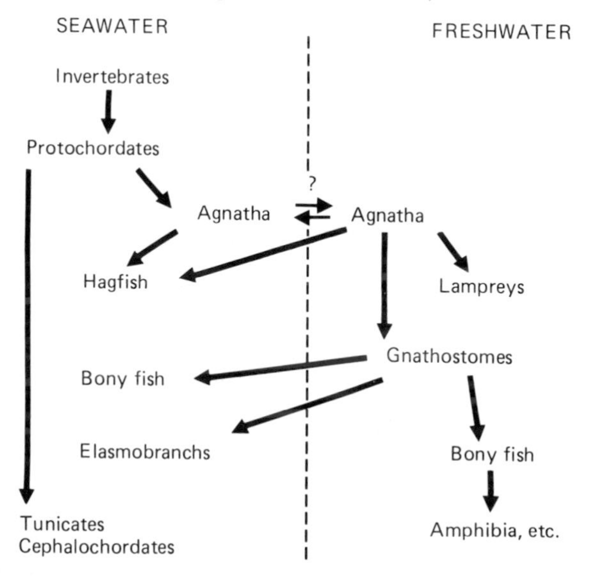

hagfish) have a body-fluid salt concentration which is approximately 30–40% of that of their ancestral protochordates (which are isosmotic to seawater, as are most non-vertebrate marine species), it appears that one of the early adaptations to life in a hypo-osmotic environment was a reduction in body-fluid concentration. Reinvasion of the marine environment by both the bony (Osteichthyes) and cartilaginous (Chondrichthyes) fishes meant that now these groups were distinctly hypo-osmotic to the surrounding seawater. Therefore, this evolutionary sequence of movement from seawater to freshwater, and back into seawater (Fig. 1), dictated that both marine and freshwater species of gilled vertebrates face substantial osmotic and ionic differentials across their branchial epithelia.

Thus, freshwater fish and amphibia faced (and still face) a net influx of water and efflux of salts, while marine fish must cope with a net efflux of water and influx of salts. In the past fifty years, investigations into the physiology of the control of these net osmotic and ionic movements have moved from determinations of the sites of control to the actual transport processes involved in the maintenance of relatively constant internal water and salt concentrations in the face of these gradients across the branchial epithelium. In the past few years there has been an explosion of review works on various aspects of fish osmoregulation and the papers of Maetz (1974), Potts (1977), Kirschner (1977, 1979) and Evans (1979, 1980*a,b,c*) should be consulted for more exhaustive reviews of this subject. (The literature on gill transport in the amphibia is quite scanty and will not be discussed in this review. The review article by Alvarado (1979) should be consulted.) What has been found is that fish use the kidney in osmoregulation, but, contrary to the situation in terrestrial vertebrates, the dominant ionoregulatory control sites are extrarenal. In freshwater, the kidney (and urinary bladder, when present) excrete large volumes of diluted urine to offset the osmotic gain of water across the branchial epithelium, and the renal loss of ions by diffusion (sodium and chloride ions are the best studied) is balanced by transport systems in the gill which are able to extract sodium and chloride ions from freshwater. In seawater the vast majority of fish ingest the medium in order to balance the osmotic loss of water and compensate for the diffusional (and oral) influx of salts by branchial extrusion mechanisms. Urinary excretion of sodium chloride by marine teleosts is quite small. The marine chondrichthyean fish have 'chosen' an alternative means of control by retaining urea in the body fluids, thus raising their concentrations to a level which is slightly hyperosmotic to seawater. This produces a slight net influx of water which is presumably used to produce the urine. Nevertheless, the marine chondrichthyean fish face a net influx of salts (predominantly sodium chloride) which must be excreted. Interestingly, these fish apparently use two pathways for excess salt excretion; one via the rectal gland (emptying into the rectum) and the other via the branchial epithelium. The relative roles of these two systems are the subject of some debate (see Evans,

1979). In summary, the fish branchial epithelium is both the site where the osmoregulatory problems which these species face arise, and the dominant site of the control mechanisms which have evolved to maintain body-fluid constancy. In recent years it has become apparent that some of the branchial ionic transport systems also play a dominant role in fish nitrogen excretion and acid–base balance (see reviews by Cameron, 1978 and Heisler, 1980). This connection between ion regulation, nitrogen excretion and acid–base regulation will become apparent later in this review.

Freshwater fishes

August Krogh (1939) was the first to demonstrate that the gills of freshwater fishes are able to extract sodium and chloride ions from freshwater independently of each other. He proposed that, in order to maintain some sort of electroneutrality across the branchial epithelium, sodium ion uptake was probably coupled with the efflux of another cation (the ammonium ion, for instance) and chloride ion uptake probably took place in exchange for blood base (or bicarbonate ions). This proposition was placed on a firm experimental footing some twenty-five years later when Maetz & Garcia Romeu (1964) demonstrated that injection of ammonium or bicarbonate ions stimulated, respectively, the influx of sodium or chloride ions into the goldfish (*Carassius auratus*), and addition of these ions to the external bath inhibited the influx of sodium or chloride ions. Thus, the uptakes were independent and coupled to appropriate blood counterions. The proposition for the chloride/bicarbonate exchange has remained unchallenged and the most recent evidence (De Renzis, 1975) indicates a fairly strict coupling between the influx of chloride ions and the efflux of base (presumably bicarbonate ions). In fact, De Renzis & Maetz (1973) found that the pH of the blood of goldfish maintained in sodium sulphate solutions increased, as did the chloride influx measured after these fish were transferred to sodium chloride solutions. They hypothesised that the blood pH rose because chloride/bicarbonate exchange was not possible in the sodium sulphate solutions, and that the chloride/bicarbonate exchange was stimulated in sodium chloride solutions because of the induced alkalosis. Unfortunately, there is no chemical means of separating the bicarbonate efflux from the hydroxyl efflux in order to determine which base is the counter-ion for chloride movement. However, blood bicarbonate levels are of the order of 10 mmol l^{-1} while, given a blood (or cell) pH of approximately 7.6, the concentrations of hydroxyl ions are less than 1 mmol l^{-1}. Therefore, unless the transport system has a much greater (ten-fold) affinity for hydroxyl ions, it seems most likely that chloride/bicarbonate exchange predominates.

The concept of sodium/ammonium ion exchange has been more carefully examined. de Vooys (1968) found that ammonia efflux (one cannot chemically differentiate free ammonia (NH_3) from the ammonium ion (NH_4^+)) from the carp (*Cyprinus carpio*) did not decline in sodium-free freshwater. This implied that the ammonia efflux was not coupled to external sodium. In addition, Kerstetter, Kirschner & Rafuse (1970) found that changing the rate of uptake of sodium by the rainbow trout (*Salmo gairdneri*), by altering the amount of sodium ions in the external bath, affected the efflux of acid but did not alter the efflux of ammonia. They argued that Maetz & Garcia Romeu's data could be explained by changes in the pH of the blood or the freshwater after the addition of ammonia and that a sodium/proton exchange was the most important uptake system for sodium. Their proposition was supported by subsequent work (Kirschner, Greenwald & Kerstetter, 1973) which showed that the sodium transport inhibitor amiloride inhibited sodium uptake and both acid and ammonia efflux. This possibility of a sodium uptake carrier coupled to either protons or ammonium ions (or the sum of protons and ammonium ions) was supported by Maetz's finding that sodium influx into the goldfish is best correlated with the sum of the net acid and ammonia efflux (Maetz, 1973). More recently Payan (1978), using an isolated, perfused trout (*S. gairdneri*) head preparation, has clearly shown that sodium influx is linked to ammonia efflux. Interestingly, his data show that not all of the ammonia efflux is dependent upon external sodium; in fact, the efflux of ammonia in sodium-free solutions is some 75% of that in normal (1 mmol l^{-1} Na^+) freshwater. Thus, not all the ammonia efflux runs through the sodium/ammonium ion exchange. It is possible that the residual efflux of ammonia occurs via simple diffusion of free ammonia across the branchial epithelium; however, recent data (Goldstein, *Gill nitrogen excretion,* in this volume and Goldstein, Claiborne & Evans, 1982) indicate that all of the ammonia traverses the gill as ammonium ions (see below). Therefore, why utilise sodium/ammonium exchange to extrude ammonia when simple diffusion of ammonium ions is possible? One might argue that it is an artefact of the need for the electroneutrality of the sodium-uptake system, with the assumption that sodium uptake is the primary 'need' for the system. However, recent evidence (see below) indicates that a sodium/ammonium exchange is present in the branchial epithelium of marine fish, where sodium uptake is certainly not the primary 'need' for a sodium/ammonium exchange. The question remains unanswered.

Very little direct evidence exists to support a sodium/proton exchange in the branchial epithelium of freshwater fishes, despite the logic of its presence. Cameron's (1976) finding that hypercapnic arctic graylings (*Thymallus arcticus*) displayed an increased sodium influx could, theoretically, be criticised on the grounds that hypercapnia (and its resulting fall in blood pH) could have a multitude of effects on branchial cell metabolism which could, in turn, affect extraction of sodium from freshwater. However, logically, most of these effects (presumably mediated through the pH sensitivity of transport or permeability-limiting proteins) would be negative, and therefore inhibit, rather than stimulate, sodium uptake. In addition, the finding of Kirschner *et al.* (1973) that amiloride inhibited both sodium influx and proton efflux could be criticised because the acid efflux was calculated as the sum of ammonia (presumed to be free ammonia) and acid efflux. Their calculation assumed that all of the ammonia entering the experimental solutions surrounding the gill epithelium immediately reacted with protons from water to form ammonium ions, thereby reducing the concentration of protons and the apparent net efflux of acid. Since their ammonia effluxes were higher than their net acid effluxes, it is clear that their experimental fish were actually excreting base, rather than acid, during the course of some of their experiments. In addition, recent evidence (Goldstein, *Gill nitrogen excretion,* in this volume and Goldstein, Claiborne & Evans, 1982) indicates that all the ammonia crosses the fish branchial epithelium as ammonium ions. These data represent the sum total of the direct evidence for a sodium/proton exchange in freshwater fishes. They can only be treated as preliminary and presumptive. The concept of a sodium/proton exchange in the branchial epithelium is logical (in terms of the metabolic needs of the animal for ion and acid–base regulation), but, like so many models in biology, we need more direct investigations of the phenomenon before accepting the model as fact. An obvious approach would be more careful measurements of acid efflux in order to test for its external sodium sensitivity. In addition, isolated, perfused head or gill preparations could be utilised so that careful control of irrigation and perfusion flows and pressures, as well as external and internal ionic concentrations and pH, is possible.

Despite the paucity of real data, it is probably safe to assume that freshwater fishes regulate their ionic concentrations in this hypo-osmotic salinity by extraction of sodium and chloride ions in exchange for blood ammonium ions and/or protons, and bicarbonate ions, respectively. Thus, this ionoregulatory system has the potential to function in nitrogen excretion and acid–base regulation (see below). It should be added that the data for the mechanisms of sodium chloride extraction by the branchial epithelium of freshwater fishes are derived entirely from work on teleosts. We assume that freshwater holostean (gar, bowfin), chondrostean (sturgeon) and dipnoan (lungfish) fishes possess the same transport mechanisms, but we have no data on which to support this proposition.

While it is generally agreed that the branchial epithelium of freshwater fishes carries out sodium/ammonium, sodium/hydrogen and chloride/bicarbonate ionic exchanges, the actual cellular and subcellular sites of these transport systems are unknown. One might assume that the interlamellar, mitochondria-rich chloride cells are the site of these systems, since they are the only cells in the branchial epithelium which display the subcellular characteristics shared by many cells that are known to transport ions (Berridge & Oschman, 1972). However, at least two lines of evidence indicate that cells other than the chloride cells may be the site of transport in freshwater fishes. Girard & Payan (1980), utilising the isolated, perfused, freshwater trout head found that sodium uptake into the perfusate occurred entirely via the lamellar compartment. They proposed, therefore, that sodium/ammonium and sodium/proton exchange took place across the lamellar epithelium, rather than the filamental epithelium which contains chloride cells. In addition, approximately 60% of the isolated opercular epithelium of the killifish (*Fundulus heteroclitus*) consists of chloride cells (see below) and no evidence for either a sodium/ammonium or a sodium/proton exchange has been published to date. In fact, even the tissue isolated from freshwater-acclimated individuals extrudes salt (chloride ions) rather than transporting it from the mucosal to the serosal side (Degnan, Karnaky & Zadunaisky, 1977). Thus, while it may be logical to assume that the chloride cell is the site of the active transport steps in the freshwater fish branchial epithelium, the only evidence we have for the site(s) of these steps points away from the chloride cell.

We must be equally hesitant about the subcellular localisation of the transport steps. We must remember that any transporting cell has basolateral (serosal, blood side) and apical (mucosal, medium side) membranes that divide the transport pathway into three (medium, cytoplasm and blood) compartments separated by two barriers. Once a transport system is designated, it is of great interest to determine the role of the two membranes in this process even though the actual cell involved may not yet be determined. This is possible, at least in a rather crude way, because the effects of ionic substitutions on transepithelial electric potentials and fluxes may give some insight into the localisation of transport steps, even across epithelia consisting of many different types of cells. The original model of Maetz & Garcia Romeu (1964) proposed that sodium/ammonium (we would now propose that sodium/proton also be included) exchange took place on the apical surface of the cell (Fig. 2*a*). This seemed to be the most logical site for the active step since freshwater sodium concentrations were certainly far below those proposed for the cytoplasm. In addition, the fact that amiloride inhibits both sodium influx and acid (and ammonia) efflux across the irrigated trout gill (Kirschner *et al.*, 1973) supports the conclusion that the sodium/ammonium and sodium/proton exchanges are apical, because amiloride is generally thought to act by blocking the entry step of sodium uptake in a variety of epithelial tissues. A basolateral sodium/potassium exchange mediating the movement of sodium into the blood was indicated by Shuttleworth & Freeman's (1974) finding that potassium-free Ringer's solution perfusing an eel (*Anguilla dieffenbachii*) gill inhibited net sodium influx. In addition, it has been shown that ouabain inhibits the sodium influx into a perfused trout gill (Richards & Fromm, 1970). Since injection of the carbonic anhydrase inhibitor acetazolamide inhibited sodium and/or chloride influx in the goldfish (Maetz & Garcia Romeu, 1964) and the trout (Kerstetter *et al.*, 1970), it appears that intracellular carbonic anhydrase must produce exchange ions (protons and/or ammonium (by interaction between protons and free ammonia) and bicarbonate) which are necessary for apical sodium and chloride uptake (Fig. 2*a*).

This model for the localization of various transport steps is still generally accepted (Evans, 1979; Kirschner, 1979); however, more recent evidence indicates that alternatives may exist, as shown diagrammatically in Fig. 2*b*. There is a distinct possibility that apical sodium uptake does not take place against an electrochemical gradient, i.e. is not active and requiring energy. It has recently

Fig. 2. Proposed cellular mechanisms of sodium and chloride ion uptake by the gills of freshwater fishes. See text for details. ((*a*) Modified from Maetz & Garcia Romeu, 1964. (*b*) Revised model based on more recent evidence.)

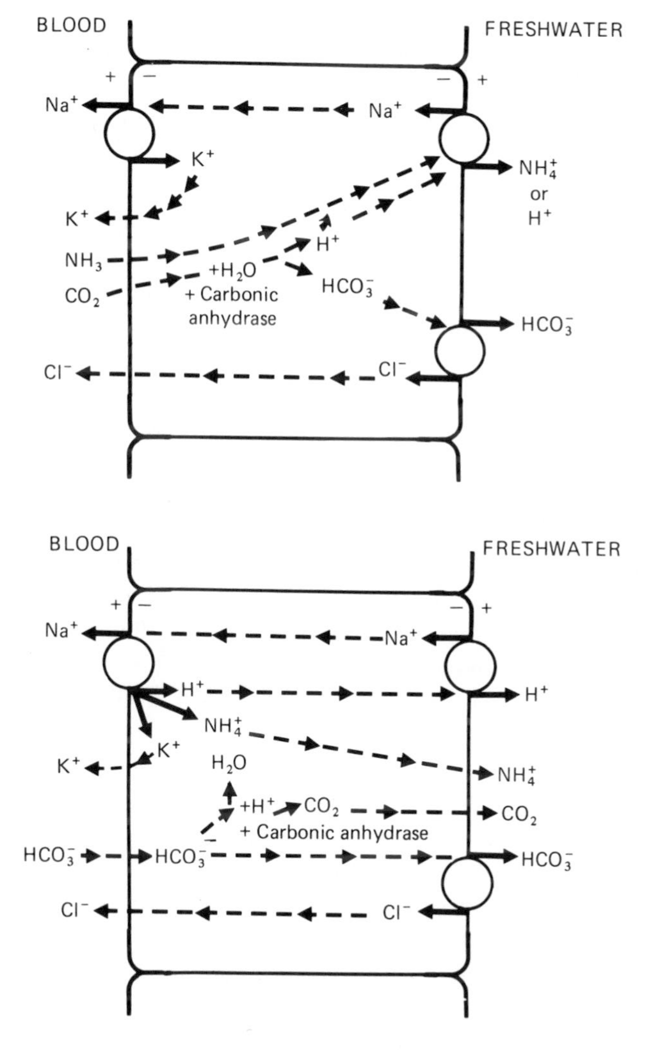

been shown that, despite large chemical gradients favouring a loss of sodium from cells, the cells of at least frog skin are sufficiently electronegative to produce electrochemical gradients which favour passive sodium influx from even quite dilute bathing solutions (Helman, 1979). Our tentative model couples this sodium influx to an extrusion of protons, because assuming that the pH of both cells and freshwater is approximately 7.5, the large electronegativity of the cell means that protons must be actively extruded across the apical membrane. Whether a direct input of energy is needed for this transport system, or whether the movement of sodium down a slight electrochemical gradient is sufficient to move protons up an opposing electrochemical gradient remains to be determined. Apical chloride/biocarbonate exchange remains a good estimate since, given the chloride concentration of freshwater and the internal negativity of the cell, it is obvious that energy must be expended in order to move chloride ions into the cell from the freshwater. However, there is some evidence which supports the conclusion that the cellular bicarbonate comes from blood bicarbonate that diffuses across the basolateral plasma membrane rather than via production (using carbonic anhydrase) inside the cytoplasm. This proposition is supported by recent data which indicate that carbon dioxide excretion by the perfused trout gill is not inhibited in erythrocyte-free Ringer's solutions, is correlated with the bicarbonate concentration of the perfusate and is inhibited by acetazolamide (Haswell, Randall & Perry, 1980; Randall, Burggren, Farrell & Haswell, 1981). These authors propose that, since over 90% of blood carbon dioxide is carried as bicarbonate and erythrocytes are apparently not necessary for the conversion of bicarbonate ions into free carbon dioxide (via erythrocyte carbonic anhydrase, as occurs in mammals), bicarbonate ions must diffuse across the basolateral plasma membrane. However, intracellular carbonic anhydrase is necessary (hence the acetazolamide sensitivity) for the dehydration of bicarbonate ions and resulting production of cellular carbon dioxide, which diffuses across the apical membrane into the freshwater. Since there is some evidence for an apical chloride/bicarbonate exchange (see above) we have suggested (Fig. 2*b*) that some of the cellular bicarbonate ions remain and are exchanged for chloride ions. The idea of excretion of carbon dioxide as the gas rather than bicarbonate ions, is supported by Cameron's (1976) finding that bicarbonate excretion by the arctic char (assuming a chloride/bicarbonate exchange ratio of nearly one) could account for only some 3% of the measured excretion of carbon dioxide. However, it is important to note that Dejours (1969) found that carbon dioxide excretion by the goldfish was stopped in chloride-free freshwater solutions. Whether these represent species differences, or experimental protocol differences, remains to be seen.

The proposition that both sodium/ammonium and sodium/proton exchanges are basolateral, and occur via the same carrier system as the sodium/potassium exchange, is supported by at least three lines of evidence. Kerstetter & Keeler

(1976) measured the sodium influx into isolated, previously perfused trout gills and found that addition of ammonium chloride stimulated sodium influx, but alteration of the pH (and therefore the amount of free-base ammonia) of the perfusate had no effect. They concluded that ammonia must enter the cells as ammonium ions, and therefore presumably via a basolateral sodium/ammonium exchange. Goldstein, Claiborne & Evans (1982), using the isolated, perfused heads of two marine teleosts (see section on marine fish for data indicating that a sodium/ammonium exchange takes place in seawater fish), showed that alteration of the concentration of free ammonia (by changing the pH of the perfusate) did not affect ammonia efflux, but changing the concentration of ammonium ions in the perfusate did. They concluded that all the ammonia efflux across the gills occurred in the ammonium-ion form and non-ionic diffusion of free ammonia did not take place. In addition, in an extension of these studies using the toadfish (*Opsanus beta*), Claiborne, Evans & Goldstein (1982) have found that addition of ouabain (0.2 mmol l^{-1}) to the perfusate inhibits the ammonia efflux by some 51% and addition of potassium ions (2.6 mmol l^{-1}) to a potassium-free Ringer's solution perfusate significantly inhibits (by 24%) ammonia efflux. Payan (1978) also found that ouabain inhibited the ammonia efflux from the perfused trout head. The ouabain effect indicates that ammonia efflux occurs either via a mechanism which includes a basolateral sodium- and potassium-activated ATPase or via an apical extrusion system that is secondarily controlled by the activity of a basolateral sodium/potassium exchange. In the latter case one could propose that ouabain inhibited sodium efflux from the cell into the blood and the resulting increase in intracellular sodium competed with intracellular ammonium ions for an apical sodium/ammonium ionic exchange system. However, the fact that addition of potassium ions to the perfusate inhibited ammonia efflux, supports the conclusion that ammonia entry into the cell occurs through a sodium- and potassium-activated ATPase system. In this case, if the sodium/ammonium exchange were apical, stimulation of basolateral sodium/potassium exchange (by adding potassium ions to the perfusate) would have reduced the intracellular concentration of sodium and therefore (if intracellular sodium and ammonium ions were competing for the same transport site) ammonia efflux should have increased, rather than decreased. The proposition that the sodium/ammonium exchange may run through the sodium- and potassium-activated ATPase is supported by the finding that ammonium ions have an even greater affinity for the ATPase than do potassium ions (Mallery, 1979). It is possible that the intracellular ammonium ions are extruded by the same apical system as the one that we propose for the sodium/proton exchange. However, one would have to propose that it is not sensitive to the intracellular concentration of sodium. In addition, since freshwater levels of ammonium ions are presumably quite low, there may be an electrochemical gradient favouring the passive efflux of ammonium ions, despite the intracellular negativity. One must be reminded that the models in Figs. 2*a, b* are quite tentative and not mutually exclusive. They are presented as a potential framework for future thought and work.

Finally, it is interesting to examine the importance of the sodium/ammonium, sodium/proton and chloride/bicarbonate exchange mechanisms in nitrogen excretion and acid–base balance in freshwater fishes. It is obvious that these transport systems are of prime importance in ionoregulation in freshwater. Are they also major pathways for nitrogen excretion and acid–base regulation? Heisler (1980) has recently reviewed mechanisms for the regulation of pH by freshwater and marine fish, and with one exception the data fully support the proposition that these pathways are the prime sites of nitrogen and acid excretion. Cameron & Wood (1978) showed that a mixed respiratory and metabolic acid load in two species of Amazonian fish (*Hoplias malabaricus* and *Hoplerythrinus unitaenatus*) is excreted primarily via non-renal extrusion of acid and Cameron (1980) has recently shown that hypercapnic catfish (*Ictalurus punctatus*) primarily use extrarenal mechanisms to compensate for the fall in blood pH. Further, Kobayashi & Wood (1980) found that infusion of lactate into the trout resulted in an increase in renal excretion of lactate and acid, but that over 72-h period only 2% of the lactate load and 6% of the proton load was excreted renally. These data are to be contrasted with an earlier study by Wood & Caldwell (1978) which showed clearly that a fixed mineral acidosis (produced by injection of hydrochloric acid) was compensated by an increased renal excretion of acid. In fact, in this study the entire acid load was excreted in 72 h by the kidney. Thus, it appears clear that (with the possible exception of a fixed mineral-acid load in the trout) extrarenal, presumably branchial, extrusion of acid predominates over renal excretion of protons by freshwater fish. There are few recent data on the role of branchial versus renal excretion of ammonia in freshwater fishes. Smith (1929) demonstrated that branchial ammonia efflux accounted for over 90% of the total and Fromm (1963) showed that renal ammonia efflux accounted for only some 3% of the total ammonia efflux from *S. gairdneri*. Thus, it appears clear that branchial sodium/hydrogen and sodium/ammonium exchanges are the dominant pathways for the excretion of acid and ammonia by freshwater fishes.

Seawater fishes

Homer Smith (1930) was the first to demonstrate that salt secretion by marine teleost fish is predominantly extrarenal, but it was not until the advent of radioisotopes that dissection of the actual branchial mechanisms was possible. There is some controversy as to whether both sodium and chloride ions actually need to be actively transported in order to maintain ionic balance in seawater. Ions move down electrochemical gradients, rather than merely chemical gradients, so it is of paramount importance to know the transepithelial electrical potential (TEP) across a given membrane before statements about active transport can be made. In freshwater fish, the TEPs are too small (Evans, 1980*b*) to account for the extremely large ionic gradients that exist between the very dilute freshwater and the fish blood (or branchial cell cytoplasm). However, the TEP across the gills of a substantial number of marine fish is of the order of +20 to 30mV (blood relative to medium), near to the equilibrium potential for sodium (Evans, 1980*a, b*), but vastly different from the equilibrium potential for chloride. In these species, it is apparent that despite substantial ionic gradients favouring the net influx of both sodium and chloride ions across the branchial epithelium, electrical gradients counter the net influx of sodium ions, but add to the net influx of chloride ions. Simply put, this means that at least some marine teleost fish need only excrete a net amount of chloride ions in order to maintain sodium-chloride balance. Two complications to this neat conclusion remain, however. Some marine teleosts have TEPs far removed from the equilibrium potentials of either sodium or chloride ions (Evans, 1980*a*) and all marine teleosts drink seawater, which will present a measurable sodium and chloride load in addition to any net diffusional uptake which may take place. In order to maintain sodium equilibrium under these conditions, the fish should have a TEP greater (more blood-positive) than the equilibrium potential for sodium, and to date this has been found in only one species, *Sarotherodon mossambica,* the tilapia (Dharmamba, Bornancin & Maetz, 1975). Thus, one must be cautious about being too definite about which ions need to be transported across the marine fish gill, at least based upon the information from TEP versus equilibrium potential data.

The literature on marine fish gill ionic transport is substantial (especially when compared to that on freshwater fish gill ionic transport) and reviews by Maetz (1974), Maetz & Bornancin (1975), Kirschner (1977, 1979, 1980), Potts (1976, 1977) and Evans (1979, 1980*a, b, c*) should be consulted for a more complete description of the relevant data. Most of these reviews also deal with the data on freshwater fish branchial transport, but a review by Maetz, Payan & De Renzis (1976) deals specifically with this subject.

When the rate of sodium and chloride ion movement across the gill of marine teleost fish is measured with radioisotopes it is readily apparent that the branchial epithelium is much more permeable to sodium and chloride ions in marine species than in freshwater species (Evans, 1979). This is presumably related to the fact that the ionic gradients across the epithelium in freshwater are distinctly greater than these gradients across the tissue in marine environments (300 : 1 versus 400 : 1000; blood:medium in freshwater and seawater respectively). Interestingly, marine teleosts are less permeable to water than freshwater teleosts (Evans, 1979) and the osmotic gradient is higher (0.013 mole fraction differential versus 0.005 mole fraction differential) in seawater. Thus it appears that teleosts are able to control passive branchial permeability to both ions and water in order to minimise net ionic and osmotic movements. One is left with the intriguing question of why, if control is possible (especially of ions), do marine teleosts maintain an exceedingly high ionic permeability of their gill tissue? This is an especially interesting question to ask when one considers that marine elasmobranchs maintain an extremely low ionic permeability, in the range of that described for freshwater teleosts (Evans, 1979). The morphological and physiological controls of these differing permeabilities to ions (and water) are unstudied.

Since many of the ionic flux studies preceded investigations of the TEP across the marine teleost fish branchial epithelium (and the knowledge that sodium may be in electrochemical equilibrium), they involved study of the kinetics of sodium transport, because the radioisotope of sodium is much easier to work with than that of chloride. Early investigations (Motais, Garcia Romeu & Maetz, 1966; Evans, 1967) demonstrated that a considerable percentage of the efflux of both sodium and chloride ions was dependent upon the presence of these ions in the external medium. The phenomenon of sodium/sodium and chloride/chloride exchange diffusion was invoked to explain these so-called trans-concentration effects (Motais *et al.*, 1966). In addition, Maetz (1969) demonstrated that sodium efflux from the flounder (*Platichthys flesus*) was dependent upon external potassium, and suggested that sodium extrusion occurred via a sodium/potassium exchange, with the sodium/sodium exchange diffusion being an artefact of the ability of seawater sodium to compete with seawater potassium (the concentration of sodium in seawater is fifty times that of potassium) for a site on the transport system. This proposition fitted nicely with concurrent biochemical data which indicated that the branchial epithelia of marine teleosts contain rather high concentrations of sodium- and potassium-activated ATPase (Jampol & Epstein, 1970) and the later finding that addition of ouabain (a specific inhibitor of this ATPase) to the external medium inhibited sodium efflux from at least two species of marine teleosts (Motais & Isaia, 1972; Evans, Mallery & Kravitz, 1973). This intellectually pleasing idea that sodium extrusion by the marine fish gill epithelium occurs via sodium- and potassium-activated ATPase can be criticised on at least two very important grounds (see reviews by Kirschner, 1979 and Evans 1979, 1980*a, b* for a more complete discussion). First, as discussed previously, measurements of the TEP across the gills of marine teleosts indicate that many species probably maintain sodium in electrochemical equilibrium. More importantly, various studies showed that alterations of the TEP could account for many of the flux changes observed under the experimental conditions used in the investigations on the presence of the sodium/sodium and sodium/potassium exchanges (e.g. Potts & Eddy, 1973; House & Maetz, 1974; Kirschner, Greenwald & Sanders, 1974). Secondly, and most critically, localisation of the supposed transport enzyme (sodium- and potassium-activated ATPase) has shown unequivocally that it is bound to the basolateral, rather than apical, plasma membranes of the chloride cell (Karnaky, Kinter & Stirling, 1976; Hootman & Philpott, 1979; Philpott, 1980). In addition, injection of ouabain into one species of marine fish inhibited sodium efflux to a greater extent than had been described after external addition (Silva, Solomon, Spokes & Epstein, 1977).

It is important to stress two things at this point. Experiments to determine the localisation of the sodium- and potassium-activated ATPase have been performed on only three species, but leave little doubt that it is basolateral. The TEP data are more equivocal, with some species of teleosts maintaining TEPs distinctly below the equilibrium potential for sodium and some species displaying flux changes (under the conditions of ionic substitutions in seawater) which cannot be accounted for by changes in the TEP (Evans, Carrier & Bogan, 1974; Maetz & Pic, 1975; Evans & Cooper, 1976; Fletcher, 1978). Thus the most critical criticism of the proposition for apical sodium/potassium exchange concerns the localisation of the enzyme, and the conflicting kinetic and TEP studies do not allow us to make a definitive statement regarding either the presence or mechanisms of active extrusion of sodium across the branchial epithelium of marine teleost fishes. It is clear that more species should be examined, preferably using histochemical, electrical and kinetic methods of investigation.

Despite the uncertainty of active sodium extrusion, it is apparent that sodium efflux is somehow coupled to chloride efflux, contrary to the situation in freshwater teleosts, where the sodium and chloride uptake systems are basically uncoupled (Fig. 2). This proposition is derived from the findings that injection of thiocyanate into the eel (*Anguilla anguilla*) and the mullet (*Mugil capito*) was followed by an increase in blood sodium and chloride ion concentrations. In addition, both species displayed a potassium-stimulated sodium and chloride efflux, both of which were inhibited by the injection of thiocyanate (Epstein, Maetz & de Renzis, 1973; Maetz & Pic, 1975). Further, injection of ouabain into the eel not only inhibited sodium- and potassium-activated ATPase and sodium ion efflux, it also inhibited chloride ion efflux (Silva *et al.*, 1977). However, some of this ouabain sensitivity may have been secondary to cardiovascular effects since ouabain injection also inhibited the efflux of water by some 40%. Unfortunately, very few studies have investigated the mechanisms of chloride extrusion directly, despite the fact that it clearly has to be actively transported. Recently, Kormanik & Evans (1979) showed that chloride extrusion by the toadfish (*Opsanus beta*) is sensitive to the external concentration of bicarbonate, independent of any changes in the TEP or the seawater pH. They suggested that chloride extrusion might occur via chloride/bicarbonate exchange, an inversion, possibly, of the freshwater chloride uptake system.

The preceding discussion outlines the majority of the data on the mechanisms of sodium and chloride extrusion that have been derived from studies on the intact organism, the marine teleost fish. It is clear that definitive statements (from these data) about the mechanisms of sodium and chloride extrusion are difficult, owing, in large part, to the inherent problems of studying transport across an epithelium which is still attached to an intact organism. We are simply unable to perform the kind of ionic substitution and electrochemical analyses which would answer most of the more intriguing questions about extrusion of sodium and chloride ions by the gill tissue of marine fishes.

Fortunately, in the past four years it has been possible to apply more rigorous techniques to this study because it has been found that at least four species of fish possess areas of their non-branchial epithelium (actually skin) which contain high densities of chloride cells and can be excised and examined in a more biophysical manner than is possible with intact fish. These tissues were first described in the inner opercular epithelium of the killifish (*Fundulus heteroclitus*) by Karnaky, Degnan & Zadunaisky (1977) and have since been recognised in the chin skin of the blenny (*Gillichthys mirabilis*) (Marshall, 1977; Marshall & Bern, 1980), the opercular skin of tilapia (*S. mossambicus*) (Foskett, Turner, Logsdon & Bern, 1979) and the opercular skin of *F. grandis* (Krasny & Zadunaisky, 1978; Krasny & Evans, 1980). The best studied tissue is that of *F. heteroclitus*, but preliminary studies on the other tissues have corroborated the data for the killifish. When mounted as a flat sheet in a chamber which allows control of mucosal and serosal solutions, as well as the electrical potential generated by the tissue, the isolated opercular epithelium of *F. heteroclitus* is able to generate a serosal-(blood) positive TEP of some 18mV when bathed on both sides with fish Ringer's solution. Since no chemical gradients are present it is clear that the tissue is transporting negative charge outward or positive charge inward, i.e. electrogenic active transport is taking place. The current necessary to abolish this TEP (the short-circuit current, I_{sc}) is of the order of 140 μA cm^{-2}. Most importantly, the I_{sc} is identical to the current carried by the net flux of chloride ions outward, no net fluxes of sodium ions are found (Degnan *et al.*, 1977). In addition, under open-circuited conditions (where no I_{sc} is applied and the tissue is bathed by seawater on the mucosal side and fish Ringer's solution on the inside), the tissue develops a greater (32mV) serosal-positive TEP (secondary, presumably, to diffusion potentials as sodium and chloride ions diffuse across cation-selective barriers) and continues to transport chloride ions against the electrochemical gradient, while sodium ions are passively distributed (Degnan & Zadunaisky, 1979). Thus it appears that this tissue, which contains a high concentration of the chloride cells thought to be the major transport pathway for salt extrusion in seawater teleost fish, actively extrudes chloride ions, but maintains sodium ions in passive equilibrium secondary to a TEP generated by the active efflux of the chloride ions as well as cation-selective permeability barriers.

Based upon their finding that injection of ouabain into *A. rostrata* inhibited both sodium and chloride efflux, their knowledge of the TEPs across the gills of many marine teleosts and the histochemical localisation of the sodium- and potassium-activated ATPase, Silva *et al.* (1977) proposed a model for the extrusion of chloride and sodium ions across the marine teleost gill epithelium which is shown diagrammatically in Fig. 3. It is important to note that this is not a model unique to this tissue (Frizzell, Field & Schultz, 1979). Recent evidence from experiments with the isolated opercular epithelium supports this model to a considerable degree. For instance, it has been shown that addition of ouabain to, or removal of sodium from, the serosal side inhibits the I_{sc} and the efflux of chloride ions (two complementary measurements of the active transport of chloride) (Degnan *et al.*, 1977; Degnan & Zadunaisky, 1980). In recent years other approaches *in vitro* have been taken with fish branchial tissue. Perfused heads and gills have generally been utilised to study haemodynamics and sodium and chloride extraction systems in freshwater (see Girard & Payan, 1980; Claiborne & Evans, 1981, for relevant references). However, it has recently been shown that the perfused gill arch of the pinfish (*Lagodon rhomboides*) actively extrudes chloride and that this efflux is at least partially dependent upon the concentration of sodium in the perfusate (Farmer & Evans, 1981). In addition it has recently been found (J. B. Claiborne, unpublished observations) that the chloride efflux from the perfused head of *O. beta* is dependent upon perfusate potassium (and inhibited by ouabain in the perfusate) and hence the presence of sodium/potassium exchange at the basolateral border. Both of these studies support the 'Silva model' for branchial ionic extrusion in marine teleosts.

We know virtually nothing about any presumed ionic extrusion systems in the

Fig. 3. Proposed cellular mechanisms of sodium and chloride ion extrusion by marine teleost fishes. See text for details. (Adapted from Silva *et al.*, 1977.)

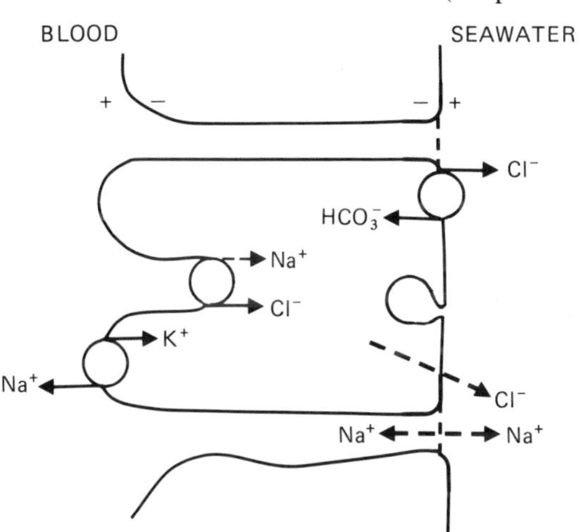

branchial cells of marine elasmobranchs; indeed there is some controversy over whether they are even necessary, given the presence of the rectal gland (see Evans, 1979 for a more complete discussion). However, since the rectal gland is apparently not vital to continued osmoregulation in at least two species (Burger, 1965; Chan, Phillips & Chester Jones, 1967) and the rectal gland efflux is less than 50% of the total sodium efflux (Burger & Tosteson, 1966; Maetz & Lahlou, 1966; Horowicz & Burger, 1968), it appears that branchial extrusion mechanisms must be present. The TEP across the few elasmobranchs which have been examined is only a few millivolts (blood negative to seawater) so it appears that both sodium and chloride ions are maintained out of electrochemical equilibrium (Evans, 1980*a*). The ionic permeabilities of elasmobranchs are extremely low (Evans, 1979) and the resulting slow rates of ionic efflux make kinetic analyses of sodium and chloride movements difficult. Preliminary data from our laboratory indicate that at least three species display a sodium efflux which is not dependent upon external sodium or potassium ions (D. H. Evans & G. A. Kormanik, unpublished observations; Evans, 1980*a*). It would be quite interesting to investigate various parameters of sodium and chloride efflux from a perfused elasmobranch head or gill arch so that more careful control of ionic substitutions is possible.

In the past few years it has become apparent that, contrary to what logic dictates, marine teleost and elasmobranch fishes actually extract sodium ions from seawater in exchange for blood/cell ammonium ions and protons (Evans, 1975*a*, 1977, 1980*a*, *b*, *c*). Indeed, more data now exist supporting the presence of at least a sodium/ammonium exchange in marine fishes than in freshwater teleosts, where this system is ionically appropriate (Evans, 1980*a*). It therefore appears that teleosts and elasmobranchs utilise the same branchial pathways for nitrogen excretion as freshwater fishes and one could argue (see below) that this supports the conclusion that marine fishes (both teleosts and elasmobranchs) arose from freshwater ancestors and have retained this system because of the overriding need for nitrogen (and possibly proton) excretion (Evans, 1975*b*). The relevant question of the magnitude of the resulting sodium load has only been applied to a single species (*O. beta*) and it appears that the net influx via the sodium/ammonium exchange is significant (twice the oral ingestion load) but not substantial (Evans, 1980*d*). However, one should remember that *O. beta* is one of the few species of fish that may face a net diffusional influx of sodium ions (its TEP is inside-negative) (Evans & Cooper, 1976), hence the relative importance of sodium influx via sodium/ammonium exchange may be much smaller in this species than in those forms which presumably maintain sodium in electrochemical equilibrium. It is only in the past two years that we have been able to test properly for the presence of a sodium/proton exchange in marine fish. We have recently found that both *O. beta* and the marine elasmobranch *Squalus* *acanthias* excrete measurable quantities of acid across their branchial epithelia (usually only after the production of acidosis via hypercapnia or injection of acid), the rate of excretion being entirely dependent upon the presence of external sodium ions (Evans, 1982). We have been quite surprised that acid (and ammonia) efflux from the marine hagfish (*Myxine glutinosa*) is also dependent upon external (seawater) sodium (Evans, 1980*e*). This finding indicates that sodium/ammonium and sodium/proton exchanges are functioning in the branchial epithelium of a modern representative of the only group of living vertebrates which has apparently never entered freshwater during its evolution (Hardisty, 1979). One must therefore conclude that, either the hagfishes have had a period in freshwater or the sodium/ammonium and sodium/proton exchanges (so important in ionoregulation in freshwater) arose in marine ancestors of the earliest vertebrates which subsequently entered the freshwater environment. Thus, at least the sodium ionoregulatory system necessary for survival in freshwater actually arose in seawater in order to facilitate the extrusion of the unwanted nitrogen and acid. Therefore it appears quite clear that nitrogen and acid excretion are the raison d'être of the sodium/ammonium and sodium/proton exchanges, rather than sodium balance. We have no direct evidence in support of the parallel system of chloride/bicarbonate exchange in marine fish; however, our experiments indicate that, when sodium/proton exchange is stopped by placing *O. beta*, *S. acanthias* or *M. glutinosa* into sodium-free artificial seawater, all three species cause the external medium to become more alkaline (Evans, 1980*e*, 1982). The intriguing proposal that this represents a chloride/bicarbonate exchange which is normally masked by a dominant sodium/proton exchange needs to be investigated.

Finally, how important are the sodium/ammonium and sodium/proton exchanges in the regulation of the pH of marine fish blood and the excretion of nitrogen? The reader is referred to Heisler's (1980) recent review, but it is pertinent to review briefly what has been published to date. Randall, Heisler & Drees (1976) demonstrated that the gills of hypercapnic spotted dogfish (*Scyliorhinus stellaris*) are able to extract bicarbonate ions from the medium (or else secrete protons, since one cannot differentiate between these two alternatives) and Heisler, Weitz & Weitz (1976) showed that this uptake was important in compensating for the fall in blood pH produced by the hypercapnia. Renal acid efflux was found to be negligible by Heisler *et al.* (1976) and it has been recently shown that the renal efflux of acid (and ammonia) can only account for 7% of a mineral acid load injected into *S. acanthias* (King & Goldstein, 1979). We have recently found that *S. acanthias* responds to either hypercapnic or mineral acid loading by branchial acid extrusion, as does the teleost, *O. beta,* with renal efflux being negligible (Evans, 1982). Thus it is clear that the branchial sodium/ammonium and sodium/proton exchanges are the dominant routes for the excretion of ammonia and acid by marine teleosts and elasmobranchs.

The studies from the author's laboratory have been supported by various grants from the National Science Foundation, most recently PCM77–09915 and PCM80–03866. Our recent research at the Mount Desert Island Biological Laboratory was supported by NSF grant PCM77–26790 and NIH Biomedical Research Support Grant SO7–RR–05764 to that institution.

References

Alvarado, R. H. (1979). Amphibians. In *Comparative Physiology of Osmoregulation in Animals,* vol. 1, ed. G. M. O. Maloiy, pp. 261–303. London: Academic Press.

Berridge, M. J. & Oschman, J. L. (1972). *Transporting Epithelia.* New York and London: Academic Press.

Burger, J. W. (1965). Roles of the rectal gland and the kidneys in salt and water excretion in the spiny dogfish. *Physiological Zoology,* **38,** 191–6.

Burger, J. W. & Tosteson, D. C. (1966). Sodium influx and efflux in the spiny dogfish (*Squalus acanthias*). *Comparative Biochemistry and Physiology,* **19,** 649–53.

Cameron, J. N. (1976). Branchial ion uptake in arctic grayling: resting values and effects of acid–base disturbance. *Journal of Experimental Biology,* **64,** 711–25.

Cameron, J. N. (1978). Regulation of blood pH in teleost fish. *Respiration Physiology,* **33,** 129–44.

Cameron, J. N. (1980). Body fluid pools, kidney function, and acid–base regulation in the freshwater catfish *Ictalurus punctatus. Journal of Experimental Biology,* **86,** 171–85.

Cameron, J. N. & Wood, C. M. (1978). Renal function and acid–base regulation in two Amazonian Erythrinid fishes: *Hoplias malabaricus,* a water-breather, and *Hoplerythrinus unitaenatus,* a facultative air-breather. *Canadian Journal of Zoology,* **56,** 917–30.

Chan, D. K. O., Phillips, J. G. & Chester Jones, I. (1967). Studies on electrolyte changes in the lip-shark *Hemiscyllium plagiosum* (Bennett), with special reference to hormonal influence on the rectal gland. *Comparative Biochemistry and Physiology,* **23,** 185–98.

Claiborne, J. B. & Evans, D. H. (1981). The effect of perfusion and irrigation flow rate variations on NaCl efflux from the isolated, perfused head of the marine teleost, *Myoxocephalus octodecimspinosus. Marine Biology Letters,* **3,** 123–30.

Claiborne, J. B., Evans, D. H., & Goldstein, L. (1982). Fish branchial Na^+/NH_4^+ exchange is via basolateral Na^+-K^+-activated ATPase. *Journal of Experimental Biology,* **96,** 431–4.

Degnan, K. J., Karnaky, K. J., Jr. & Zadunaisky, J. A. (1977). Active chloride transport in the in-vitro skin of a teleost (*Fundulus heteroclitus*), a gill-like epithelium rich in chloride cells. *Journal of Physiology,* **271,** 155–91.

Degnan, K. J. & Zadunaisky, J. A. (1979). Open-circuit Na^+ and Cl^- fluxes across isolated opercular epithelial from the teleost, *Fundulus heteroclitus. Journal of Physiology* **294,** 483–95.

Degnan, K. J. & Zadunaisky, J. A. (1980). Ionic contributions to the potential and current across the opercular epithelium. *American Journal of Physiology,* **238,** R231–9.

Dejours, P. (1969). Variations of CO_2 output of a fresh-water teleost upon change of the ionic composition of the water. *Journal of Physiology*, **202**, 113*P*.

De Renzis, G. (1975). The branchial chloride pump in the goldfish *Carassius auratus*: Relationship between Cl^-/HCO_3^- and Cl^-/Cl^- exchanges and the effect of thiocyanate. *Journal of Experimental Biology*, **63**, 587–602.

De Renzis, G. & Maetz, J. (1973). Studies on the mechanisms of the chloride absorption by the goldfish gill: relation with acid–base regulation. *Journal of Experimental Biology*, **59**, 339–58.

Dharmamba, M., Bornancin, M. & Maetz, J. (1975). Environmental salinity and sodium chloride exchanges across the gill of *Tilapia mossambica*. *Journal de Physiologie* (Paris), **70**, 627–36.

Epstein, F. H., Maetz, J. & de Renzis, G. (1973). On the active transport of chloride by the teleost gill. Inhibition by thiocyanate. *American Journal of Physiology*, **224**, 1195–9.

Evans, D. H. (1967). Sodium, chloride and water balance of the intertidal teleost, *Xiphister atropurpureus*. III. The roles of simple diffusion, exchanged diffusion, osmoisis and active transport. *Journal of Experimental Biology*, **47**, 525–34.

Evans, D. H. (1975*a*). The effects of various external cations and sodium transport inhibitors on sodium uptake by the sailfin molly, *Poecilia latipinna*, acclimated to sea water. *Journal of Comparative Physiology*, **96**, 111–15.

Evans, D. H. (1975*b*). Ionic exchange mechanisms in fish gills. *Comparative Biochemistry and Physiology*, **51A**, 491–5.

Evans, D. H. (1977). Further evidence for Na/NH_4 exchange in marine teleost fish. *Journal of Experimental Biology*, **70**, 213–220.

Evans, D. H. (1979). Fish. In *Comparative Physiology of Osmoregulation in Animals*, vol. 1, ed. G. M. O. Maloiy, pp. 305–90. London: Academic Press.

Evans, D. H. (1980*a*). Kinetic studies of ion transport by fish gill epithelium. *American Journal of Physiology*, **238**, R224–30.

Evans, D. H. (1980*b*). Salt transport mechanisms in branchial epithelia. In *Animals and Environmental Fitness*, ed. R. Gilles, pp. 61–78. London: Pergamon Press.

Evans, D. H. (1980*c*). Osmotic and ionic regulation by freshwater and marine fish. In *Environmental Physiology of Fishes*, ed. M. A. Ali, pp. 93–122. New York: Plenum Press.

Evans, D. H. (1980*d*). Na/NH_4 exchange in the marine teleost, *Opsanus beta*; stoichiometry and role in Na balance. In *Epithelial Transport in the Lower Vertebrates*, ed. B. Lahlou, pp. 197–205. Cambridge University Press.

Evans, D. H. (1980*e*). Acid and ammonia excretion by *Squalus acanthias* and *Myxine glutinosa*: effect of hypercapnia, acid injection and Na-free sea water. *Bulletin of the Mount Desert Island Biological Laboratory*, **20**, 60–3.

Evans, D. H. (1982). Mechanisms of acid extrusion of two marine fishes: The teleost, *Opsanus beta*, and the elasmobranch *Squalus acanthias*. *Journal of Experimental Biology*, in press.

Evans, D. H., Carrier, J. C. & Bogan, M. B. (1974). The effect of external potassium ions on the electrical potential measured across the gills of the teleost, *Dormitator maculatus*. *Journal of Experimental Biology*, **61**, 277–83.

Evans, D. H. & Cooper, K. (1976). The presence of Na/Na and Na/K exchange in sodium extrusion by three species of fish. *Nature*, **259**, 241–2.

Evans, D. H., Mallery, C. H. & Kravitz, L. (1973). Sodium extrusion by a fish acclimated to sea water: physiological and biochemical description of Na-for-K exchange system. *Journal of Experimental Biology*, **58**, 627–36.

Farmer, L. L. & Evans, D. H. (1981). Chloride extrusion in the isolated perfused teleost gill. *Journal of Comparative Physiology,* **141,** 471–6.
Fletcher, C. R. (1978). Osmotic and ionic regulation in the cod (*Gadus callarias* L.). II. Salt balance. *Journal of Comparative Physiology,* **124,** 157–68.
Foskett, J. K., Turner, T., Logsdon, C. & Bern, H. (1979). Electrical correlates of the chloride-cell development in subopercular membrane of the tilapia *Sarotherodon mossambicus* transferred to sea water. *American Zoologist,* **19,** 995a.
Frizzell, R. A., Field, M. & Schultz, S. G. (1979). Sodium-coupled chloride transport by epithelial tissues. *American Journal of Physiology,* **236,** F1–8
Fromm, R. O. (1963). Studies on renal and extra-renal excretion in a freshwater teleost, *Salmo gairdneri. Comparative Biochemistry and Physiology,* **10,** 121–8.
Girard, J. P. & Payan, P. (1980). Ion exchanges through respiratory and chloride cells in freshwater- and seawater-adapted teleosteans. *American Journal of Physiology,* **238,** R260–8.
Goldstein, L., Claiborne, J. B. & Evans, D. H. (1982). Ammonia excretion by the gills of two marine fish: The importance of NH_4^+ permeance. *Journal of Experimental Zoology,* **219,** 395–7.
Hardisty, M. W. (1979). *Biology of Cyclostomes.* 428 pp. London: Chapman and Hall.
Haswell, M. S., Randall, D. J. & Perry, S. F. (1980). Fish gill carbonic anhydrase: acid–base regulation or salt transport? *American Journal of Physiology,* **238,** R240–5.
Heisler, N. (1980). Regulation of the acid–base status in fishes. In *Environmental Physiology of Fishes,* ed. M. A. Ali, pp. 123–62. New York: Plenum Press.
Heisler, N., Weitz, H. & Weitz, A. M. (1976). Hypercapnia and resultant bicarbonate transfer processes in an elasmobranch fish (*Scyliorhinus stellaris*). *European Bulletin of Respiratory Physiopathology,* **12,** 77–85.
Helman, S. I. (1979). Electrochemical potentials in frog skin: inferences for electrical and mechanistic models. *Federation Proceedings,* **38,** 2743–50.
Hootman, S. R. & Philpott, C. W. (1979). Ultracytochemical localization of Na^+, K^+-activated ATPase in chloride cells from the gills of a euryhaline teleost. *Anatomical Record,* **193,** 99–129.
Horowicz, P. & Burger, L. W. (1968). Unidirectional fluxes of Na in the spiny dogfish, *Squalus acanthias. American Journal of Physiology,* **214,** 635–42.
House, C. R. & Maetz, J. (1974). On the electrical gradient across the gill of the sea water-adapted eel. *Comparative Biochemistry and Physiology,* **47A,** 917–24.
Jampol, L. M. & Epstein, F. H. (1970). Sodium–potassium-activated adenosine triphosphatase and osmotic regulation by fishes. *American Journal of Physiology,* **218,** 607–11.
Karnaky, K. L., Jr, Degnan, K. J. & Zadunaisky, J. A. (1977). Chloride transport across isolated opercular epithelium of killifish: a membrane rich in chloride cells. *Science* (N.Y.), **195,** 203–5.
Karnaky, K. L., Jr, Kinter, W. B. & Stirling, C. E. (1976). Teleost chloride cell. II. Autoradiographic localization of gill Na, K-ATPase in killifish, *Fundulus heteroclitus,* adapted to low and high salinity environments. *Journal of Cellular Biology,* **70,** 157–77.
Kerstetter, T. H. & Keeler, M. (1976). On the interaction of NH_4^+ and Na^+

fluxes in the isolated trout gill. *Journal of Experimental Biology,* **64,** 517–27.

Kerstetter, T. H., Kirschner, L. B. & Rafuse, D. D. (1970). On the mechanisms of sodium ion transport by the irrigated gills of rainbow trout (*Salmo gairdneri*). *Journal of General Physiology,* **56,** 342–59.

King, P. A. & Goldstein, L. (1979). Dogfish (*Squalus acanthias*) renal ammonia response to an acid load. *Bulletin of the Mount Desert Island Biological Laboratory,* **19,** 77–80.

Kirschner, L. B. (1977). The sodium chloride excreting cells in marine vertebrates. In *Transport of Ions and Water in Animals,* ed. B. L. Gupta, P. B. Moreton, J. L. Oschmand, & B. W. Wall, pp. 427–52. London and New York: Academic Press.

Kirschner, L. B. (1979). Control Mechanisms in Crustaceans and Fishes. In *Osmoregulation in Animals,* ed. P. Gilles, pp. 157–222. New York: Wiley Interscience.

Kirschner, L. B. (1980). Comparison of vertebrate salt-excreting organs. *American Journal of Physiology,* **238,** R219–23.

Kirschner, L. B., Greenwald, L. & Kerstetter, T. H. (1973). Effect of amiloride on sodium transport across body surfaces of fresh water animals. *American Journal of Physiology,* **224,** 832–7.

Kirschner, L. B., Greenwald, L. & Sanders, M. (1974). On the mechanism of sodium extrusion across the irrigated gill of sea water-adapted rainbow trout (*Salmo gairdneri*). *Journal of General Physiology,* **64,** 148–65.

Kobayashi, K. A. & Wood, C. M. (1980). The response of the kidney of the freshwater rainbow trout to true metabolic acidosis. *Journal of Experimental Biology,* **84,** 227–44.

Kormanik, G. A. & Evans, D. H. (1979). HCO_3-stimulated Cl efflux in the gulf toadfish acclimated to sea water. *Journal of Experimental Zoology,* **208,** 13–16.

Krasny, E. J., Jr & Evans, D. H. (1980). Effects of catecholamines on active Cl^- secretion by the opercular epithelium of *Fundulus grandis. The Physiologist,* **23,** 63a.

Krasny, E., Jr & Zadunaisky, J. A. (1978). Ion transport properties of the isolated opercular epithelium of *Fundulus grandis. Bulletin of the Mount Desert Island Biological Laboratory,* **18,** 117–18.

Krogh, A. (1939). *Osmotic Regulation in Aquatic Animals.* Cambridge University Press. Dover Publications reprint, 242 pp.

Lutz, P. L. (1975). Adaptive and evolutionary aspects of the ionic content of fishes. *Copeia,* **1975,** 369–73.

Maetz, J. (1969). Sea water teleosts: evidence for a sodium–potassium exchange in the branchial sodium-excreting pump. *Science* N.Y., **166,** 613–15.

Maetz, J. (1973). Na^+/NH_4^+, Na^+/H^+ exchanges and NH_3 movement across the gill of *Carassius auratus. Journal of Experimental Biology,* **58,** 255–75.

Maetz, J. (1974). Aspects of adaptation to hypo-osmotic and hyper-osmotic environments. In *Biochemical and Biophysical Perspectives in Marine Biology,* vol. 1, ed. D. C. Malins & J. R. Sargent, pp. 1–167. London and New York: Academic Press.

Maetz, J. & Bornancin, M. (1975). Biochemical and biophysical aspects of salt secretion by chloride cells in teleosts. International Symposium on Excretion. *Fortschritte der Zoologie,* **28,** Heft 2/3, 322–62.

Maetz, J. & Garcia Romeu, F. (1964). The mechanism of sodium and chloride uptake by the gills of a fresh water fish, *Carassius auratus* II. Evidence for

NH_4^+/Na^+ and HCO_3^-/Cl^- exchanges. *Journal of General Physiology*, **47**, 1209–27.

Maetz, J. & Lahlou, B. (1966). Les echanges de sodium et de chlore chez un Elasmobranche, *Scyliorhinus*, mesures a l'aide des isotopes ^{24}Na et ^{36}Cl. *Journal de Physiologie* (Paris), **58**, 249.

Maetz, J. & Pic, P. (1975). New evidence for a Na/K and Na/Na exchange carrier linked with the Cl pump in the gill of *Mugil capito* in sea water. *Journal of Comparative Physiology*, **102**, 85–100.

Maetz, J., Payan, P. & De Renzis, G. (1976). Controversial aspects of ionic uptake in fresh water animals. In *Perspectives in Experimental Biology*, vol. 1, *Zoology*, ed. P. Spencer Davies, pp. 77–92. Oxford and New York: Pergamon Press.

Mallery, C. H. (1979). Ammonium stimulated properties of K-dependent ATPase in *Opsanus beta*, a teleost with an NH_4^+/Na^+-exchange pump. *American Zoologist*, **19**, 944a.

Marshall, W. S. (1977). Transepithelial potential and short-circuit current across the isolated skin of *Gillichthys mirabilis* (Teleostei: Gobiidae), acclimated to 5% and 100% sea water. *Journal of Comparative Physiology*, **114**, 575–8.

Marshall, W. S. & Bern, H. A. (1980). Ion transport across the isolated skin of the teleost, *Gillichthys mirabilis*. In *Epithelial Transport in the Lower Vertebrates*, ed. B. Lahlou, pp. 337–50. Cambridge University Press.

Motais, R., Garcia Romeu, F. & Maetz, J. (1966). Exchange diffusion effect and euryhalinity in Teleosts. *Journal of General Physiology*, **50**, 391–442.

Motais, R. & Isaia, J. (1972). Evidence for an effect of ouabain on the branchial sodium-excreting pump of marine teleosts: interaction between the inhibitor and external Na and K. *Journal of Experimental Biology*, **57**, 367–73.

Payan, P. (1978). A study of the Na^+/NH_4^+ exchange across the gill of the perfused head of the trout (*Salmo gairdneri*). *Journal of Comparative Physiology*, **124**, 181–8.

Philpott, C. W. (1980). Tubular system membranes of teleost chloride cells: osmotic response and transport sites. *American Journal of Physiology*, **238**, R171–84.

Potts, W. T. W. (1976). Ion transport and osmoregulation in marine fish. In *Perspectives in Experimental Biology*, vol. 1, ed. P. Spencer Davies, pp. 65–75. Oxford: Pergamon Press.

Potts, W. T. W. (1977). Fish gills. In *Transport of Ions and Water in Animals*, ed. B. L. Gupta, R. B. Moreton, J. L. Oschmand, & B. W. Wall, pp. 453–80. London and New York: Academic Press.

Potts, W. T. W. & Eddy, F. B. (1973). Gill potentials and sodium fluxes in the flounder *Platichthys flesus*. *Journal of Comparative Physiology*, **87**, 29–48.

Randall, D. J., Burggren, W. W., Farrell, A. P. & Haswell, M. S. (1981). *The Evolution of Air Breathing in Vertebrates*. 133 pp. Cambridge University Press.

Randall, D. J., Heisler, N. & Drees, F. (1976). Ventilatory response to hypercapnia in the larger spotted dogfish *Scyliorhinus stellaris*. *American Journal of Physiology*, **230**, 590–4.

Richards, B. D. & Fromm, P. O. (1970). Sodium uptake by the isolated perfused gills of rainbow trout (*Salmo gairdneri*). *Comparative Biochemistry and Physiology*, **33**, 303–10.

Shuttleworth, T. J. & Freeman, R. F. (1974). Factors affecting the net fluxes of ions in the isolated perfused gills of fresh water *Anguilla dieffenbachii*. *Journal of Comparative Physiology*, **94**, 297–307.

Silva, P., Solomon, R., Spokes, K. & Epstein, F. H. (1977*a*). Ouabain inhibition of gill Na–K–ATPase: Relationship to active chloride transport. *Journal of Experimental Zoology*, **199**, 419–27.
Smith, H. W. (1929). The excretion of ammonia and urea by the gills of fish *Journal of Biological Chemistry*, **81**, 727–42.
Smith, H. W. (1930). The absorption and excretion of water and salts by marine teleosts. *American Journal of Physiology*, **93**, 480–505.
Vooys, G. G. N. de (1968). Formation and excretion of ammonia in Teleostei. I. Excretion of ammonia through the gills. *Archives Internationales de Physiologie et de Biochemie*, **76**, 268–72.
Wood, C. M. & Caldwell, F. H. (1978). Renal regulation of acid–base balance in a freshwater fish. *Journal of Experimental Zoology*, **305**, 301–7.

DAVID RANDALL

Blood flow through gills

Gill blood flow and the anatomy of the branchial circulation have been extensively investigated over the past two hundred years, especially in the last two decades (Monro, 1785; Hyrtl, 1838; Müller, 1839; Riess, 1881; Steen & Kruysse, 1964; Dornesco & Miscalenco, 1969; Hughes & Morgan, 1973; Laurent & Dunel, 1976, 1980). The essential components of the system were described by Monro (1785). He injected mercury into the gill vasculature of a fish and noticed first its passage through the secondary lamellae, and then its appearance in the body of the filament, flowing back towards the heart. He compared the lamellar flow to the pulmonary circulation of the lung, and the recurrent filament flow to the bronchial circulation. This is still a reasonable comparison in that the lamellar circuit is involved in gas transfer and the capillaries supplying the filament sinus probably have a nutritive function. This comparison, however, obscures the fact that the gills, unlike the lungs, are very much involved in osmotic and ionic regulation as well as gas transfer.

Blood leaving the heart is forced through the secondary lamellae and is then collected in the dorsal aorta for distribution throughout the body. Small capillaries arise, mostly from the efferent filament artery, but also the arch efferent, the filament afferent artery and, in rare instances, the basal portions of the secondary lamellae, and connect with a series of filament sinuses (Boland & Olson, 1979; Laurent & Dunel, 1980). These sinuses are joined by lymphatic vessels to form a large venolymphatic drainage which returns to the heart. The general pattern of the microvascular organisation is similar in all fish, with variations of a quantitative rather than qualitative nature. The size of the gills and the number, size and shape of the lamellae are very variable, as is the extent of the filament sinus and the nature and number of input capillaries, whether all of these arise from the efferent side of the gill circulation or some arise from the afferent side (Grigg & Read, 1971; Wright, 1973; Vogel, Vogel & Pfautsch, 1976; Cooke, 1980; Cooke & Campbell, 1980). The extent to which the lamellae are buried in the body of the filament also is variable (Kuhn & Koecke, 1956). Air-breathing fish, for instance, often have reduced gills with small buried lamellae that have large vascular channels (B. J. Gannon, personal communication) and a thick epithelium to prevent oxygen loss to hypoxic water when air breathing (Randall, Burggren, Farrell & Haswell, 1981*a*). The microcirculation in fish with fused gills is essentially similar to that of other fish, except in tuna where there is a more linear array of pillar cells forming more obvious channels for blood flow through the lamellae (Muir, 1970). These fish have fused lamellae and filaments in order to maintain the gill dimensions either at high water flow during ram ventilation (e.g. tuna; Muir, 1970) or during air exposure (e.g. *Amia*; Daxboeck, Barnard & Randall, 1981).

Elasmobranch gills, although having a somewhat different general structure (Grigg, 1970), have a microvascular organisation which is similar to that of teleosts. There are, however, some important differences. Elasmobranch gills have an extensive series of corpora cavernosa connected to the afferent filament artery. This system presumably is at a pressure close to ventral aortic pressure and could act either as a hydraulic skeleton (Cooke, 1980) or perhaps to smooth the blood pulse coming from the heart (Wright, 1973). The former suggestion seems more likely, especially as elasmobranch gill filaments, unlike those of teleosts, lack any skeletal support. Another special feature of elasmobranch gills is that the secondary lamellae on the basal portion of gill filaments are covered by a canopy and are not ventilated with water (Cooke, 1980). The canopy probably exists to protect growing lamellae but also creates a non-ventilated blood shunt within the lamellar circuit. As blood normally approaches saturation after passing through the gills in the dogfish, this region of the filament is unlikely to be extensively perfused with blood.

A large number of animals other than fish possess gills, including crabs (Taylor & Greenaway, 1979), lampreys (Nakao & Uchinomiya, 1978) and dipnoi (Laurent, Delaney & Fishman, 1978). Some information exists on changes in blood pressure and flow in elasmobranchs (Kent & Peirce, 1978), lungfish (Johansen, Lenfant & Hanson, 1968) and crabs (Burggren, McMahon & Costerton, 1974) but the teleost gill is the most extensively studied. The subsequent discussion is restricted to the effects of changes in pressure and flow on the gill circulation in teleost fish. These results, however, may have a general qualitative applicability to vascular function in the gills of other animal groups because of the similarities in structure at the level of the microvascular circulation.

Teleostean branchial vascular anatomy

The general features of the gill circulation in a teleost fish are illustrated in Fig. 1. The gill vascular bed constitutes about 20 to 40% of the total vascular resistance of the fish. The ventral aorta divides to produce arch vessels, which in turn supply filament arteries which then supply the respiratory units of the gills, the secondary lamellae. The afferent and efferent lamellar arterioles arising from the filament arteries are the major sites of resistance to blood flow in the gills. The input pressures to all secondary lamellae are similar because the filament arteries are large, and, although tapering at the distal end, have reduced blood flow towards the tip as blood flows into successive lamellae (Farrell, 1979). The diameter of lamellar arterioles decreases with distance along the filament in lingcod (Farrell, 1980) and therefore the distal lamellae have a higher resistance to flow than more basal structures. Thus basal lamellae will be preferentially perfused. Additionally, there is more smooth muscle around more distal lamellar arterioles in trout, and perhaps many other fish.

The venolymphatic circulation in the body of the filament receives blood from the efferent filament artery. The connecting capillaries are relatively few in number, have a small diameter, are often long and tortuous, and therefore have a high resistance. Many fish also have a venolymphatic supply arising from the afferent filament artery (Vogel *et al.*, 1976; Dunel & Laurent, 1977; Vogel, 1978). The central sinus of the venolymphatic system has pocket valves which presumably allow the drainage of fluid from extravascular spaces in the filament and lamellae (Fig. 2). Changes in blood distribution and flow in the lamellae and venolymphatic system and drainage of the extravascular spaces undoubtedly will have a marked effect on gill function.

Fig. 1. The circulation in the gill filament of a typical teleost. (From Randall, Perry & Heming, 1981*b*.)

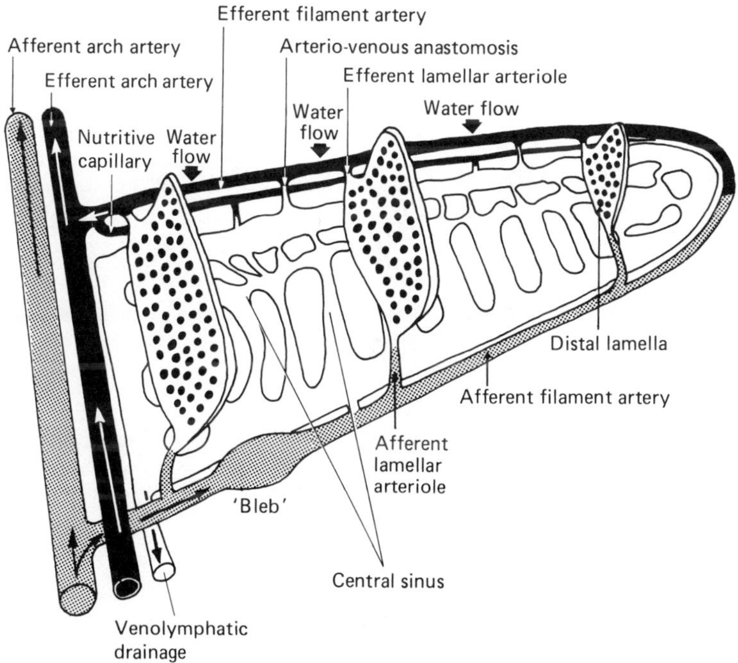

The gills are covered by epithelial, mucous and chloride cells. The lamellae are covered by flattened epithelial cells whereas the chloride cells are generally found in the region between lamellae, on the body of the filament. In freshwater fish the epithelial and chloride cells are joined by tight junctions and the epithelium has a high resistance and low permeability to ions and water (Sardet, Pisam & Maetz, 1979; Sardet, 1980). The gills of seawater fish have a lower resistance and are more permeable than those of freshwater fish, largely because the increased numbers of chloride cells seen in marine teleost gills are loosely associated with accessory cells. There is a leaky paracellular channel between these two types of cell which is probably a major pathway for sodium efflux across the gill (Karnaky, 1980). Blood pressure has been measured in a variety of fish. The secondary lamellae receive all or nearly all (some blood may flow from the afferent filament artery into the filament sinus rather than to the lamellae) the cardiac output and operate at high blood pressure. The venolymphatic circuit contains fewer erythrocytes and more white cells than the rest of the circulation (Skidmore & Tovell, 1972) and operates at low blood pressure. The leaky junctions between chloride and accessory cells in seawater fish overlay the low pressure venolymphatic circulation.

Fig. 2. Schematic cross-section through a secondary lamella and filament of a teleost gill. (From Randall *et al.*, 1981*b*.)

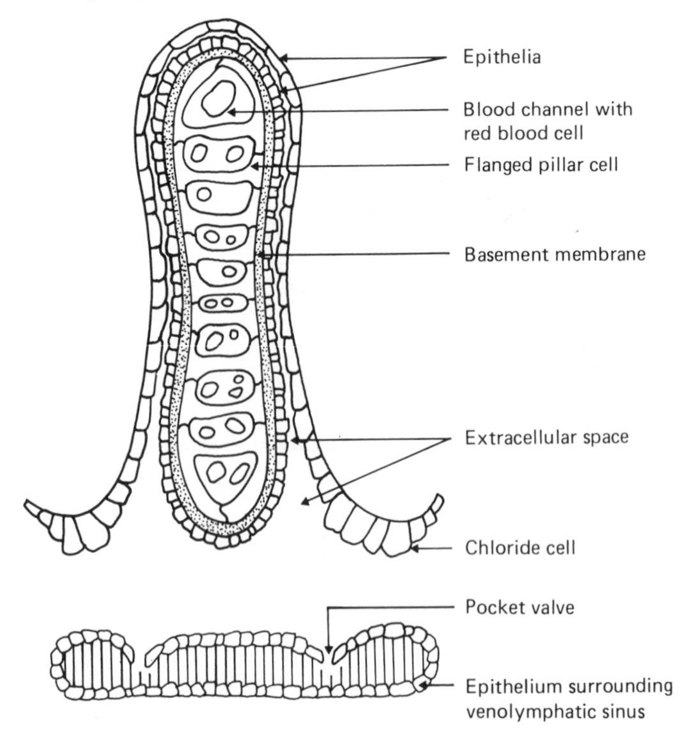

The blood in the secondary lamellae is separated from water by a pillar cell flange and two layers of epithelial cells (Fig. 2). Functionally, there appear to be two major barriers to the movement of material between blood and water; the basal (blood) barrier and the much more permeable water-facing barrier (Isaia, Girard & Payan, 1978*a*). The anatomical correlates of these barriers are not known but could be the external plasma membrane on the water side and the epithelial plasma membrane and associated basement layer on the blood side.

Pressure–flow relations

Cardiac output is a measure of lamellar blood flow in those fish with no, or only very few, afferent filamental connections to the filament sinus. Ventral and dorsal aortic pressures are indicative of the input and output pressures of the lamellar circulation. Dorsal aortic pressure also is indicative of the input pressure to the venolymphatic circulation as all vessels between the site of entrance of efferent artery venolymphatic capillaries and the dorsal aorta are large and of low resistance.

The relation between input and output pressure, and flow and lamellar resistance and recruitment, has been investigated using an isolated holobranch preparation from lingcod (*Ophiodon elongatus*), a species with known *in vivo* ventral and dorsal aortic pressures and flows (Farrell, 1979). In the isolated holobranch, using a constant flow of saline, a rise in outflow pressure results in a smaller rise in inflow pressure, thus the pressure drop and resistance to flow through the gills falls with increasing transmural pressure (Fig. 3). The gill vessels must therefore

Fig. 3. The effect of increasing outflow pressure, P_o, on (*a*) input pressure, P_i, and (*b*) gill resistance, ΔP_g, which at constant flow, $\dot{Q}$, is equivalent to the resistance to flow in the gills, R_g. (After Farrell, Daxboeck & Randall, 1979.)

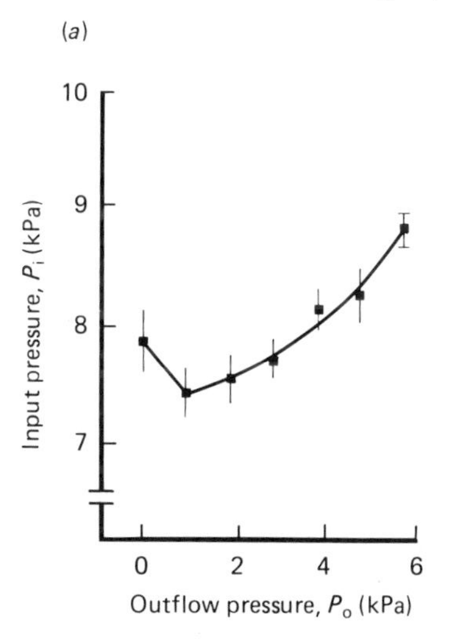

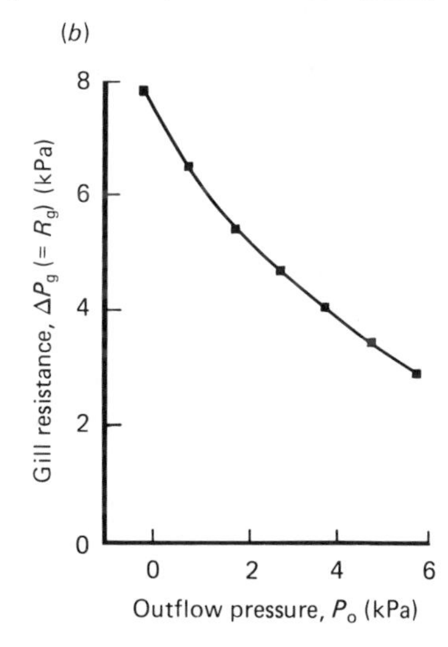

be distensible. The resistance change, however, is related to regions other than the secondary lamellae, because gill resistance was not related in any clear manner to the number of lamellae perfused (Fig. 4). This is not to imply that the lamellae do not change volume with pressure, only that any such changes are either balanced by lamellar recruitment or are relatively unimportant in determining the gill resistance to flow.

In investigations using the isolated holobranch preparations of lingcod there was a tendency for only those lamellae on the basal two-thirds of each filament to be perfused (Farrell, Daxboeck & Randall, 1979). Similar observations were made on the trout *in vivo* (Booth, 1978). If the input pressure and flow were increased in the isolated lingcod holobranch the result was lamellar recruitment. At constant flow, a decrease in pulse pressure with an increased pulse rate decreased the number of lamellae perfused, whereas an increase in pulse pressure associated with a decrease in pulse rate, but a rise in stroke volume, caused an increase in the number of lamellae perfused. Farrell *et al.* (1979) concluded that as the distal lamellar arterioles are narrower, they had a higher critical opening pressure than more proximal arterioles. Thus, in these experiments, lamellar recruitment could be explained in terms of an increased input pressure opening previously closed lamellar arterioles. Typically, once open, these lamellae will

Fig. 4. The relation between the number of lamellae perfused and gill resistance. The straight line indicates the expected relation for a simple ohmic resistance with a variable number of resistors in parallel. (After Farrell *et al.*, 1979.)

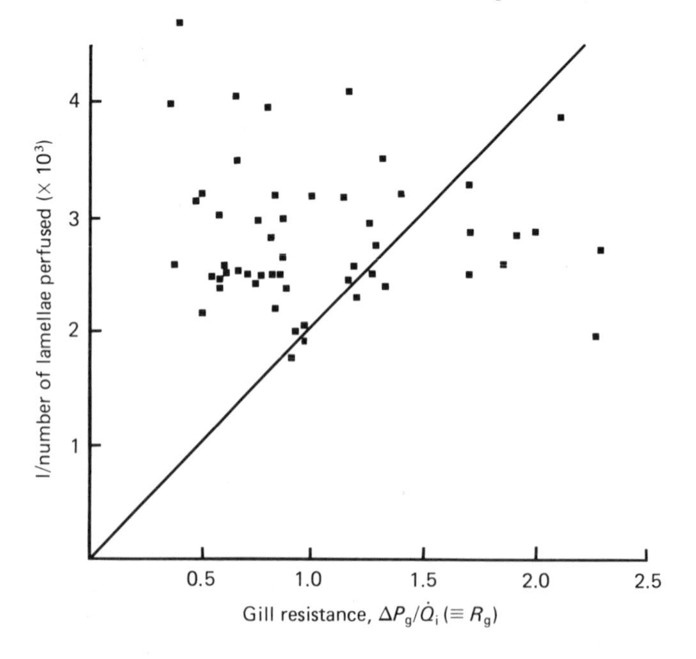

remain patent at lower pressures because the critical opening pressure undoubtedly exceeds the closing pressure by several hundred Pa. Thus the pressure increase need not be maintained in order to cause lamellar recruitment and an increase in peak pulse pressure (even if mean pressure is unchanged) could be sufficient to cause lamellar recruitment.

Changes in blood pressure also affect blood distribution within the secondary lamellae. In lingcod, the vascular space to tissue ratio of the gill lamellae is not altered but the thickness of the blood sheet is very sensitive and increases with the transmural pressure (Fig. 5). Clearly, blood flow through the secondary lamellae of lingcod and probably all fish, can be described as sheet flow (Farrell, Sobin, Randall, & Crosby, 1980) as in the lungs of man and dog (Glazier, Hughes, Maloney & West, 1969; Sobin, Fung, Tremer & Lindal, 1979). The lingcod gill has a thicker respiratory epithelium and a wider vascular sheet than mammals and so one would expect a somewhat less compliant structure. However, although less compliant than that of man, the lingcod gill vascular sheet does show about the same compliance as that observed in the blood sheet of dog lungs.

A portion of each lamella is buried in the body of the filament and exposed

Fig. 5. The relation between the thickness of the vascular sheet (closed symbols; double solid line), the vascular space to tissue ratio (open symbols; broken line) and the transmural pressure across the lamellar walls of the gills of the lingcod. The different symbols (triangle, circle and square) represent three individual fish. Standard errors and the number of observations are shown. (After Farrell *et al.*, 1980.)

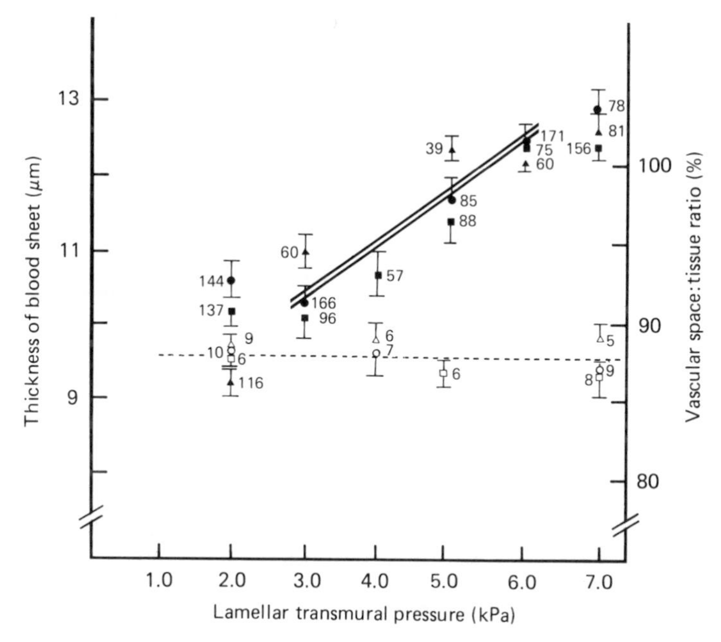

regions are more compliant than buried portions when exposed to pressure changes in the physiological range. Thus a pressure increase causes central, exposed regions to expand more than basal regions of the lamellae. As flow is proportional to h^4, where h is the thickness of the vascular sheet, this expansion results in a marked increase in flow through these central regions of the gill. Thus at lower physiological pressures, basal and marginal flow predominates but with increasing pressure central flow becomes more significant (Fig. 6). Thus a rise in transmural pressure, although increasing basal flow, causes a greater increase in flow in regions of the lamellae exposed to water flow. As basal channels are separated from the environment by a larger diffusion barrier than marginal channels, the shift toward marginal flow will increase the gas diffusing capacity of the gills.

The vascular space to tissue ratio does not change with pressure, indicating little or no change in the height or length of the lamellae. They may become more rigid and show less tendency to flap in the water flow at higher pressures, but this was not investigated in this study. The pillar cell posts do not change in length or diameter with increases in thickness of the vascular sheet; rather the flanges bulge outwards. Thus there must be some thinning of the pillar cell flanges and the space between the pillar and epithelial cells may be reduced. Extracellular columns of collagen are embedded in an orderly array in the outer surface of the pillar cells (Newstead, 1967) and extend into the flanges to connect with the basement membrane on either side of the blood sheet (Hughes & Wright, 1970). This collagen presumably determines the nature and limits to physiological expansion of the lamellar blood sheet. There are actomyosin fibrils associ-

Fig. 6. The effect of different transmural pressures (open circles, 5.0 kPa; closed circles, 3.0 kPa) on blood flow in different regions of the secondary lamellae of the lingcod. (Data from Farrell *et al*., 1980.)

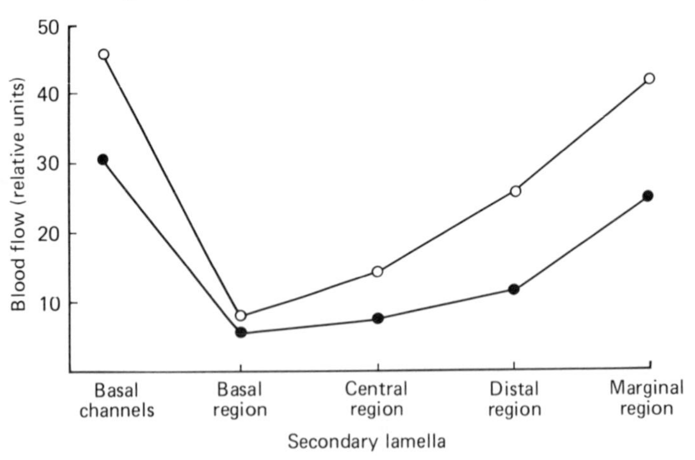

ated with the collagen (Bettex-Galland & Hughes, 1972; Wright, 1973), and although these are unlikely to be able to generate sufficient force to contract the pillar cells (G. F. Holeton, unpublished calculations), they may play some role in organising the collagen fibres along lines of stress to maximise their effectiveness in reducing expansion of the lamellar sheet. Farrell *et al.* (1980) investigated only the short-term static responses of the gills to changes in pressure. Whether the dynamic responses are the same or different is not known at this time.

Changes in pressure and flow clearly alter the perfusion of the secondary lamellae. In addition, Davie & Daxboeck (1982) have shown that pulsatility may affect the permeability of the epithelial barrier in a way similar to that reported by Isaia, Maetz & Haywood (1978*b*) for adrenaline. Not all lamellae are perfused in resting fish (Booth, 1978); the unperfused lamellae will have a reduced pulse pressure compared with perfused lamellae, and so may be less permeable to water. Recruitment of lamellae will be associated, therefore, with an increase in water flux across the gills. Both these things occur in swimming fish (Wood & Randall, 1973*b;* Jones & Randall, 1978).

Changes in blood pressure and flow also affect the venolymphatic circulation within the gills. Fluid in the filament venolymphatic sinus originates from flow in capillaries that run from both the efferent and afferent filament arteries and fluid that drains from the extracellular spaces within the gill filament. The relative contribution from these two sources to the fluid within the central sinus is not known. Blood in the sinus often has fewer erythrocytes than blood in the general circulation and a high number of white cells (Skidmore & Tovell, 1972; Booth, 1978; Soivio, Nikinmaa, Nyholm & Westman, 1981). This could be due to plasma skimming at the entrance to capillaries leading from the filament artery to the central sinus (Vogel *et al.,* 1976) as well as dilution of the blood with extracellular fluid.

Farrell (1979) was able to record a few *in vivo* venolymphatic pressures in the gills of lingcod (Fig. 7). The pressures were pulsatile and of the order of less than 1 kPa. The pressures in the extracellular spaces must be slightly higher if they are to drain into the central sinus. The pulsatility recorded in these low pressure sinuses is perhaps surprising and may be transmitted to the central sinus through the tissues rather than through the connecting but narrow blood capillaries. Thus there may be considerable mechanical interaction between the high pressure lamellar circulation and the low pressure recurrent venolymphatic system. There is certainly a close proximity of vessels to facilitate such a mechanical interaction. The venolymphatic system parallels the filament arteries and often surrounds the lamellar arterioles with doughnut-like structures. Thus pulsations in the high pressure vessels will be transmitted to the surrounding low pressure venolymphatic system and may be important in generating venolymphatic flow.

Pulsations of the pillar cell walls and other vessels may also aid in moving extracellular fluid, forcing it from the extravascular space into the central sinus. In the isolated saline-perfused trout head an increase in pulse pressure at constant input flow caused a marked increase in venolymphatic outflow (Table 1). This was due to the increase in pulsatility rather than absolute pressure because an increase in input pressure with reduced pulsatility actually reduced venolymphatic outflow. These data are from preparations in which the dorsal aorta cannula was tied in with a tourniquet around the dorsal body caudal to the heart. This tourniquet occluded flow through the subclavian arteries and smaller segmental vessels. A larger number of observations are available for preparations without the tourniquet and although the data are more variable, they illustrate the

Fig. 7. Pressures recorded from the *in vivo* venolymphatic system using a micropipette system for micropressure recordings in the lingcod gill. (From Farrell, 1979.) The upper three traces are simultaneous measures of blood pressure, the lower trace is blood flow ($\dot{Q}$) in the ventral aorta.

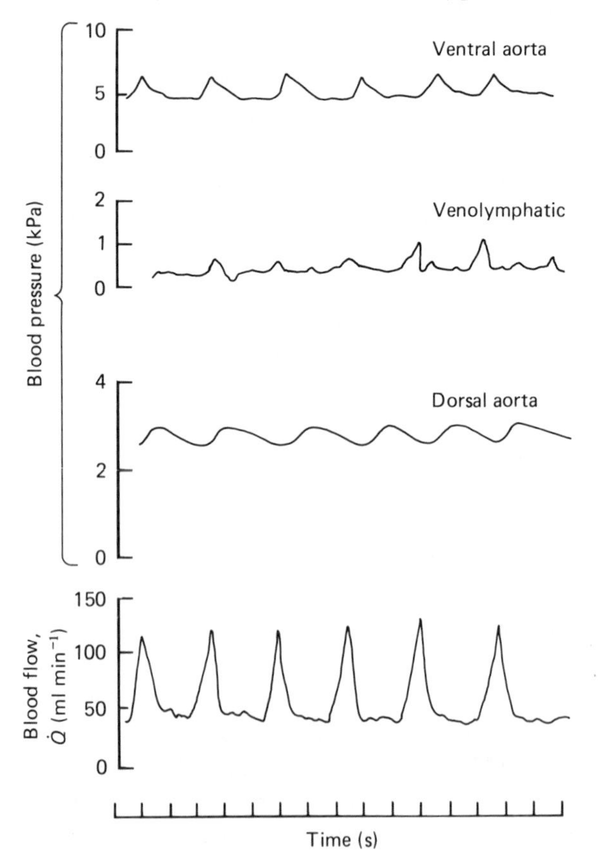

same principles. It seems probable that the effect of increased pulsatility in increasing venolymphatic flow must be either to empty the extravascular spaces or to cause flow in the venolymphatic vessels, thus preventing any fluid backup or rise in pressure in the extensively valved venolymphatic system.

Changes in pressure in the buccal cavity may also be important for emptying the venolymphatic system and synchronisation of the heart rate with the mouth-closing phase of the breathing cycle (Fig. 8) may be to maximise the interaction between the pressure pulse in the blood and water in order to augment venolymphatic flow.

Pressures in the venolymphatic system are low and it is unlikely, therefore, that this system can play any role as a hydraulic skeleton in teleost gills as suggested for the central sinus of the lamprey gill by Nakao & Uchinomiya (1978). It is possible that the filament sinus can function as a fluid reservoir (Girard & Payan, 1976) and be involved in immunological reactions associated with the large number of granulocytes found in the sinus (Skidmore & Tovell, 1972). The vessels are too large to be of nutritive value, but the small capillaries supplying the central sinus probably serve this role. As a low pressure system, the filament sinus may function in the collection of fluid that has been filtered into the extracellular spaces in the gill filament and the secondary lamellae, and has exchanged

Table 1. *The effect of various input regimes on dorsal aortic and venolymphatic flow in the isolated head preparation of the trout.* (*From Daxboeck & Davie, 1982*)

	Input pressure and flow	
	Pulsatile ($n = 16$)	Non-pulsatile ($n = 12$)
Input flow (ml min^{-1})	4.79 ± 0.19	4.41 ± 0.29
Venolymphatic outflow (ml min^{-1})	1.39 ± 0.09	1.09 ± 0.11
Outflow in dorsal aorta (ml min^{-1})	1.54 ± 0.13	1.43 ± 0.22
Saline loss (%)	38.71 ± 3.08	42.26 ± 3.75
Gill resistance (ml min^{-1} cm H_2O 100 g^{-1})	14.41 ± 1.36	18.11 ± 2.85 (18.8; Wood, 1974)
Venolymphatic resistance (ml min^{-1} cm H_2O 100 g^{-1})	11.67 ± 0.98	14.97 ± 1.83

water and ions with the medium across both the epithelial and chloride cells. None of these possible functions of the filament sinus are mutually exclusive.

Physiological consequences of changes in gill blood flow

Thus changes in pressure and flow in the gills can affect the resistance to flow, the number of lamellae perfused, the distribution of blood within the lamellae, flow in the venolymphatic vessels and the permeability of the gill epithelium. The next questions are: do these cardiovascular changes occur in the intact animals and are the changes functionally significant? The answer to the first question is yes, these cardiovascular changes do occur, but the second question cannot be answered so easily. In trout, ventral aortic pressures increase sharply with the onset of exercise and then fall to levels which are still above resting values (Stevens & Randall, 1967*a;* Kiceniuk & Jones, 1977). The resistance and capacitance of the systemic circuit are reduced (Jones, Langille, Randall & Shelton, 1974) but there is an increase in cardiac output large enough to result in an increase in both dorsal and ventral aortic pressure (Fig. 9). The

Fig. 8. The effect of oxygen concentration on synchrony between the heart and breathing cycles in the rainbow trout. Upper trace: ECG, superimposed sweeps. Lower trace: superimposed sweeps recording the movements of the lower jaw. The phases of the breathing cycle are indicated in (*d*). (*a*) Normal oxygen concentration; (*b*) oxygen concentration below critical level; (*c*) oxygen concentration below critical level but following the injection of atropine into the pericardial cavity. (From Randall & Smith, 1967.)

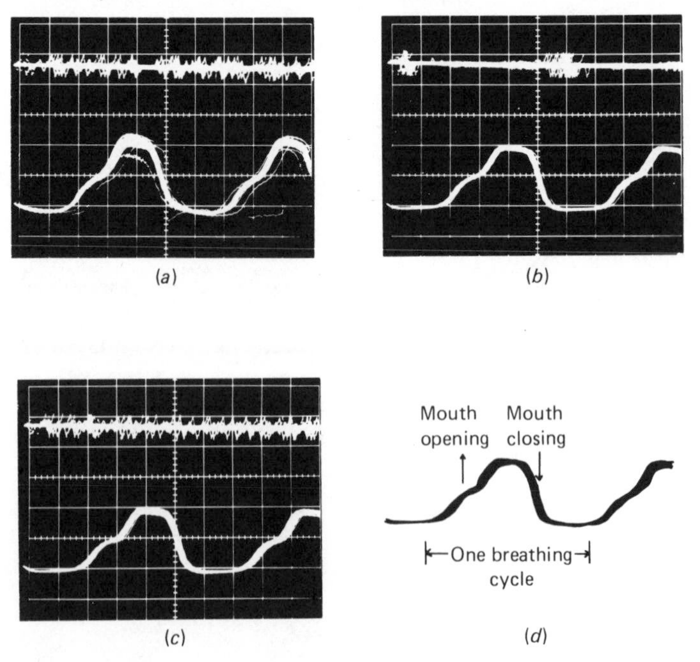

resistance to flow in the gills increases slightly during exercise in contrast to the marked fall in systemic resistance (Jones & Randall, 1978). There are increased rates of oxygen (Stevens & Randall, 1967*a,b*), sodium and water transfer (Wood & Randall, 1973*a,b*) across the gills, associated with a redistribution of blood

Fig. 9. Changes in blood pressure and flow during exercise in the trout. ((*a*) From Stevens & Randall, 1967*a*; (*b*) from Kiceniuk & Jones, 1977.)

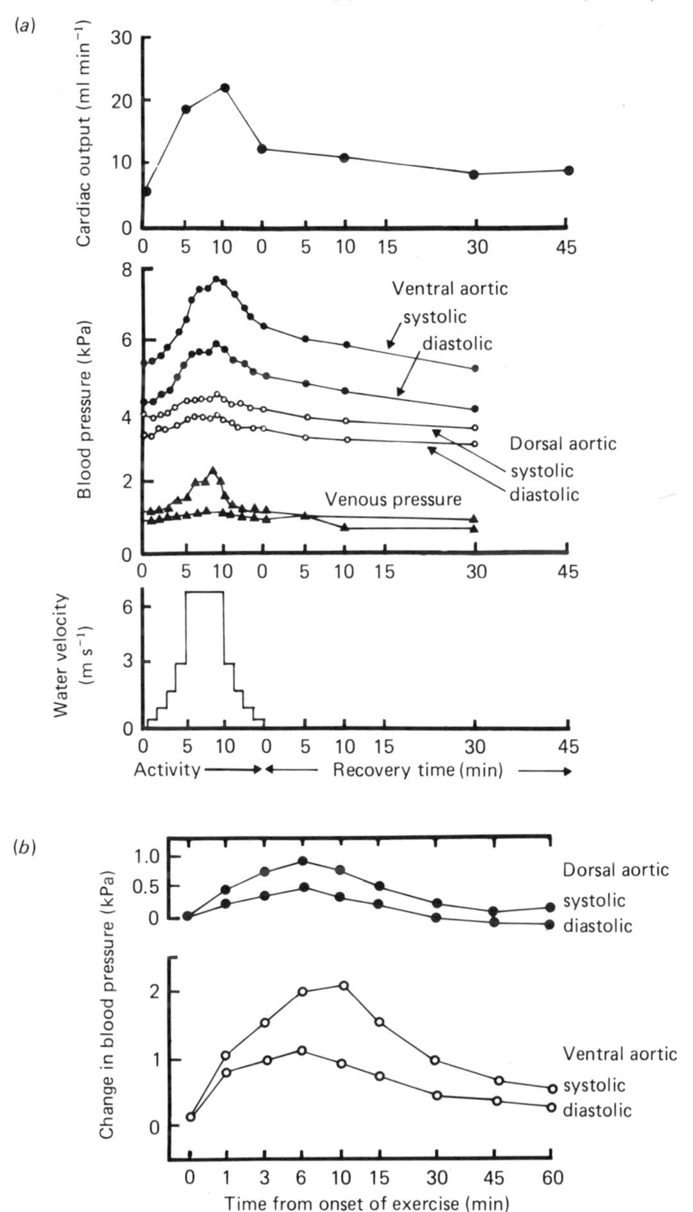

within the gills and perhaps changes in gill epithelial permeability. Endogenous catecholamine release has been implicated in these changes and has also been shown to cause lamellar recruitment (Booth, 1978; Holbert, Boland & Olsen, 1979) as well as changes in membrane permeability (Isaia *et al.*, 1978*a,b*). The changes in pressure and flow could also contribute directly to lamellar recruitment and increased gill permeability as a result of the rise in both mean pressure and pulsatility. Most probably the effects of catecholamines and increases in pressure and flow are additive in causing changes within the gill vasculature.

Hypoxia causes a bradycardia in fish, with little change in cardiac output as the decrease in heart rate is offset by an increase in stroke volume (Holeton & Randall, 1967). The rise in stroke volume is associated with a marked rise in both ventral and dorsal aortic pulse and mean pressure (Fig. 10).

The diffusing capacity of the gills has been shown to increase for carbon monoxide during hypoxia in the catfish (Fisher, Coburn & Forster, 1969). In blood-perfused head preparations changes in pressure and flow had little effect on oxygen uptake in the hypoxic trout (Daxboeck, Davie, Perry & Randall, 1982). In this preparation, oxygen uptake across the gills was dependent on blood flow and was independent of mean and pulse pressure. Arterial oxygen partial pressure (P_{aO_2}) was not affected by changes in flow indicating a perfusion rather than diffusion limited system. P_{aO_2} was also unchanged in trout exercising at levels

Fig. 10. Changes in ventral and dorsal aortic pressure in the trout during hypoxia. (From Holeton & Randall, 1967.)

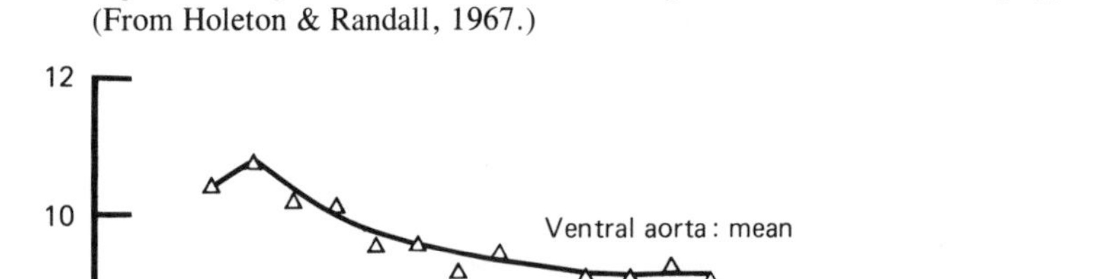

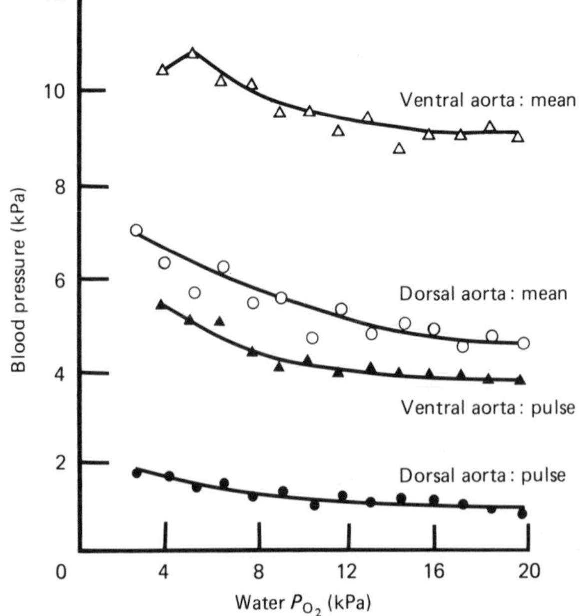

up to 95% of critical velocity (Kiceniuk & Jones, 1977). Cardiac output was calculated by Kiceniuk & Jones (1977) using the Fick principle based on measurements of oxygen uptake and arterial–venous oxygen content differences. Daxboeck & coworkers (1982) have since determined that gill tissue could account for 27% of resting oxygen uptake. Thus only 73% of the oxygen leaving the water enters the blood in a resting fish, the remainder being consumed by gill tissue. If Kiceniuk and Jones' data for cardiac output are recalculated, assuming a constant gill tissue oxygen uptake of 27% of resting oxygen uptake from water, then blood transit time through the secondary lamellae, assuming a lamellar volume of 0.7 ml kg^{-1} fish (Jones & Randall, 1978) is more than three seconds and less than one second during rest and exercise respectively (Fig. 11). These values of transit time are similar to those reported by Hughes *et al.* (1981) for the

Fig. 11. The effect of blood residence time in the secondary lamellae of a trout on arterial oxygen tension (P_{aO_2}). Lamellar volume was assumed to be 0.7 ml kg^{-1} fish (Jones & Randall, 1978). Cardiac output values were recalculated from data reported by Kiceniuk & Jones (1977) corrected for an oxygen consumption of the gill tissues of 27% of resting oxygen uptake. Venous oxygen tension (P_{vO_2}) from Kiceniuk & Jones (1977). If one assumes that all lamellae are perfused at rest then blood residence time in the lamellae is over 3 s, whereas if only 60% of lamellae are perfused then residence time is reduced to 2 s. The shaded area represents the range of P_{aO_2} found in experiments in which pressure and flow were changed in a blood-perfused trout head preparation, again assuming a lamellar volume of 0.7 ml kg^{-1} fish (Daxboeck *et al.*, 1982).

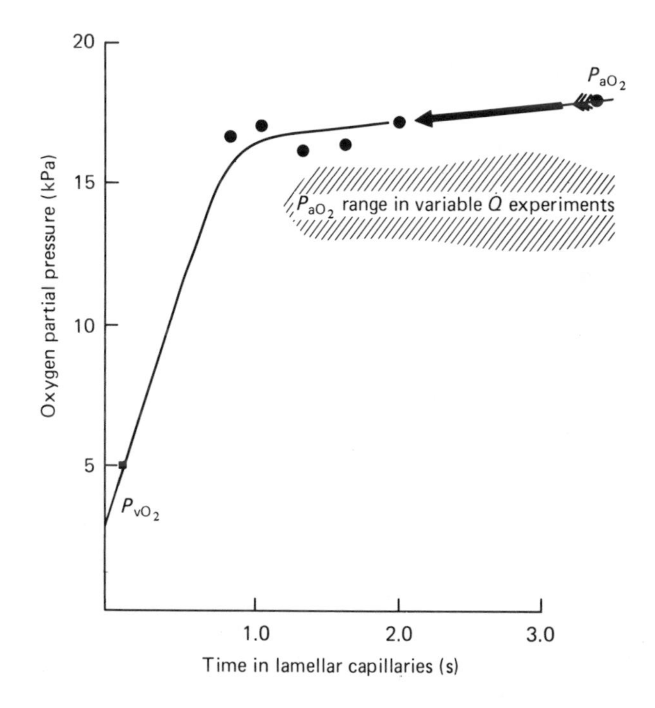

goldfish. Hughes & Koyama (1974) had shown that the time for oxygenation of blood within the gills was approximately one second in the carp and eel, that is, within the transit time of blood through the secondary lamellae. Thus P_{aO_2} is, once again, largely unaffected by transit time. If the number of lamellae perfused in trout is reduced to 60% of the total then transit time will be decreased to two seconds in the resting fish. This may reduce gill water transfer, by reducing water loss in less permeable unperfused lamellae, without affecting oxygen transfer. At high levels of exercise the redistribution of blood within the lamellae away from basal channels, the thinning of the epithelium due to expansion of the blood sheet and lamellar recruitment may augment the gill diffusing capacity for oxygen in order to maintain high levels of oxygen uptake during periods of high blood flow or during hypoxia when oxygen gradients are reduced. The changes in gill diffusing capacity, associated with changes in gill blood flow and pressure, would appear, however, to be relatively unimportant at intermediate levels of exercise and hypoxia. Under these conditions the gills appear to be perfusion limited for oxygen uptake by blood.

References

Bettex-Galland, M. & Hughes, G. M. (1972). Demonstration of a contractile actomyosin-like protein in the pillar cells of fish gills. *Experientia,* **28,** 744.

Boland, E. J. & Olson, K. R. (1979). Vascular organization of the catfish gill filament. *Cell and Tissue Research,* **198,** 487–500.

Booth, J. H. (1978). The distribution of blood flow in the gills of fish: application of a new technique to rainbow trout (*Salmo gairdneri*). *Journal of Experimental Biology,* **73,** 119–29.

Burggren, W., McMahon, B. R. & Costerton, J. W. (1974). Branchial water and bloodflow patterns and the structure of the gill of the crayfish *Procambarus clarkii*. *Canadian Journal of Zoology,* **52,** 1511–18.

Cooke, I. R. C. (1980). Functional aspects of the morphology and vascular anatomy of the gills of the Endeavour Dogfish, *Centrophorus scalpratus* (McCulloch) (Elasmobranchii: Squalidae). *Zoomorphologie,* **94,** 167–83.

Cooke, I. R. C. & Campbell, G. (1980). The vascular anatomy of the gills of the smooth toadfish *Torquiginer glaber* (Teleostei: Tetraodontidae). *Zoomorphologie,* **94,** 151–66.

Davie, P. S. & Daxboeck, C. (1982). Effect of pulse pressure on fluid exchange between the blood and tissues of trout gills. *Canadian Journal of Zoology,* in press.

Daxboeck, C. & Davie, P. S. (1982). The effects of pulse pressure on flow distribution within an isolated, saline-perfused trout head preparation. *Canadian Journal of Zoology,* in press.

Daxboeck, C., Barnard, D. K. & Randall, D. J. (1981). Functional morphology of the gills of the bowfin, *Amia calva* L., with special reference to their significance during air exposure. *Respiration Physiology,* **43,** 349–64.

Daxboeck, C., Davie, P. S., Perry, S. F. & Randall, D. J. (1982). Oxygen uptake in a spontaneously ventilating blood-perfused trout preparation. *Journal of Experimental Biology,* in press.

Dornesco, G. T. & Miscalenco, D. (1969). Etude comparative de la structure des branchies de quelques ordres de Téléostéens. *Anatomische Anzeiger,* **124,** 68–84.

Dunel, S. & Laurent, P. (1977). La vascularization branchiale chez l'Anguille: action de l'acétylcholine et de l'adrenaline sur la répartition d'une resine polymérisable dans les differents compartiments vasculaires. *Comptes Rendus de l'Académie des Sciences, Paris,* **284,** 2011–14.

Farrell, A. P. (1979). Gill blood flow in teleosts. Ph.D. Thesis, University of British Columbia, Vancouver.

Farrell, A. P. (1980). Gill morphometrics, vessel dimensions and vascular resistance in lingcod, *Ophiodon elongatus. Canadian Journal of Zoology,* **58,** 807–18.

Farrell, A. P., Daxboeck, C. & Randall, D. J. (1979). The effect of input pressure and flow on the pattern and resistance to flow in the isolated perfused gill of a teleost fish. *Journal of Comparative Physiology,* **133,** 233–40.

Farrell, A. P., Sobin, S. S., Randall, D. J. & Crosby, S. (1980). Intralamellar blood flow patterns in fish gills. *American Journal of Physiology,* **239,** R428–36.

Fisher, T. R., Coburn, R. F. & Forster, R. E. (1969). Carbon monoxide diffusing capacity in the bullhead catfish. *Journal of Applied Physiology,* **26(2),** 161–9.

Girard, J. P. & Payan, P. (1976). Effect of epinephrine on vascular space of gills and head of rainbow trout. *American Journal of Physiology,* **230,** 1555–60.

Glazier, J. B., Hughes, J. M. B., Maloney, J. E. & West, J. B. (1969). Measurements of capillary dimensions and blood volume in rapidly frozen lungs. *Journal of Applied Physiology,* **26,** 65–76.

Grigg, G. C. (1970). Water flow through the gills of Port Jackson sharks. *Journal of Experimental Biology,* **52,** 565–7.

Grigg, G. C. & Read, J. (1971). Gill function in an elasmobranch. *Zeitschrift für Vergleichende Physiologie,* **73,** 439–51.

Holbert, P. W., Boland, E. J. & Olsen, K. R. (1979). The effect of epinephrine and acetylcholine on the distribution of red cells within the gills of the channel catfish (*Ictalurus punctatus*). *Journal of Experimental Biology,* **79,** 135–46.

Holeton, G. F. & Randall, D. J. (1967). Changes in blood pressure in the rainbow trout during hypoxia. *Journal of Experimental Biology,* **46,** 297–305.

Hughes, G. N., Horimoto, M., Kikuchi, Y., Kakiuchi, Y. & Koyama, T. (1981). Blood-flow velocity in microvessels of the gill filaments of the goldfish (*Carassius auratus* L.). *Journal of Experimental Biology,* **90,** 327–31.

Hughes, G. M. & Koyama, T. (1974). Gas exchange of single red blood cells within secondary lamellae of fish gills. *Journal of Physiology,* **24b,** 82–3P.

Hughes, G. M. & Morgan, M. (1973). The structure of fish gills in relation to their respiratory function. *Biological Reviews,* **48,** 419–75.

Hughes, G. M. & Wright, D. E. (1970). A comparative study of the ultrastructure of the water/blood pathway in the secondary lamellae of teleost and elasmobranch fishes – benthic forms. *Zeitschrift für Zellforschung und Mikroskopische Anatomie,* **104,** 478–93.

Hyrtl, J. (1838). Beobachtungen aus dem Gebiethe der vergleichenden Gefäβ lehre. II. Uber den Bau der Kiemen der Fische. *Medecin Jahrbuche Osterreich Staates,* **24,** 232–48.

Isaia, J., Girard, J. P. & Payan, P. (1978*a*). Kinetic study of gill epithelial permeability to water diffusion in the freshwater trout, *Salmo gairdneri*: Effect of adrenaline. *Journal of Membrane Biology,* **41,** 337–47.

Isaia, J., Maetz, J. & Haywood, G. P. (1978*b*). Effects of epinephrine on branchial nonelectrolyte permeability in rainbow trout. *Journal of Experimental Biology,* **74,** 227–37.

Johansen, K., Lenfant, C. & Hanson, D. (1968). Cardiovascular dynamics in the lungfishes. *Zeitschrift für Vergleichende Physiologie,* **59,** 157–86.

Jones, D. R., Langille, B. L., Randall, D. J. & Shelton, G. (1974). Blood flow in the dorsal and ventral aortae of the cod (*Gadus morhua*). *American Journal of Physiology,* **266,** 90–5.

Jones, D. R. & Randall, D. J. (1978). The respiratory and circulatory systems during exercise. In *Fish Physiology,* vol. 7, ed. W. S. Hoar & D. J. Randall, pp. 425–501. New York: Academic Press.

Karnaky, K. J., Jr. (1980). Ion-secreting epithelia: chloride cells in the head region of *Fundulus heteroclitus. American Journal of Physiology,* **238,** R185–98.

Kent, B. & Peirce, E. C. II. (1978). Cardiovascular responses to changes in blood gases in dogfish shark, *Squalus acanthias. Comparative Biochemistry and Physiology,* **60C,** 37–44.

Kiceniuk, J. W. & Jones, D. R. (1977). The oxygen transport system in trout (*Salmo gairdneri*) during sustained exercise. *Journal of Experimental Biology,* **69,** 247–60.

Kuhn, O. & Koecke, H. V. (1956). Histologische und zytologische Veränderungen der Fischkieme nach Einwirkung im Wasser enthaltener schädigender Substanzen. *Zeitschrift für Zellforschung und Mikroskopische Anatomie,* **43,** 611–43.

Laurent, P., Delaney, R. G. & Fishman, A. P. (1978). The vasculature of the gills in the aquatic and aestivating lungfish *Protopterus aethiopicus. Journal of Morphology,* **156,** 173–208.

Laurent, P. & Dunel, S. (1976). Functional organization of the teleost gill. I. Blood pathways. *Acta Zoologica* (*Stockholm.*), **57,** 189–209.

Laurent, P. & Dunel, S. (1980). Morphology of gill epithelia in fish. *American Journal of Physiology,* **238,** R147–59.

Monro, A. (1785). *The Structure and physiology of fishes explained and compared with those of man and other animals.* C. Elliot: Edinburgh.

Muir, B. S. (1970). Contribution to the study of blood pathways in teleost gills. *Copeia,* **1970,** 19–28.

Müller, J. (1839). Vergleichende Anatomie der Myxinoiden. III. Uber das Gefäßsystem. *Abhandlung Akademie Wissenschaften Berlin,* 175–303.

Nakao, T. & Uchinomiya, K. (1978). A study on the blood vascular system of the lamprey gill filament. *American Journal of Anatomy,* **151,** 239–64.

Newstead, J. D. (1967). Fine structure of the respiratory lamellae of teleostean gills. *Zeitschrift für Zellforschung und Mikroskopische Anatomie,* **79,** 396–428.

Randall, D. J., Burggren, W. W., Farrell, A. P. & Haswell, M. S. (1981*a*). *The Evolution of Air Breathing in Vertebrates.* Cambridge University Press.

Randall, D. J., Perry, S. F. & Heming, T. (1982*b*). Gas transfer in the gills of salmonid fish. *Comparative Biochemistry and Physiology,* in press.

Randall, D. J. & Smith, J. C. (1967). The regulation of cardiac activity in fish in a hypoxic environment. *Physiological Zoology,* **40,** 104–13.

Riess, J. A. (1881). Der Bau der Kiemenblätter bei den Knochenfischen. *Archives Naturgeschaft,* **47,** 518–50.

Sardet, C. (1980). Freeze fracture of the gill epithelium of euryhaline teleost fish. *American Journal of Physiology,* **238,** R207–12.
Sardet, C., Pisam, M. & Maetz, J. (1979). The surface of epithelium of teleostean fish gills. Cellular and junctional adaptations of the chloride cell in relation to salt adaptations. *Journal of Cell Biology,* **80,** 96–117.
Skidmore, J. F. & Tovell, P. W. A. (1972). Toxic effects of zinc sulfate on the gills of rainbow trout. *Water Research,* **6,** 217–30.
Sobin, S. S., Fung, Y. C., Tremer, H. M. & Lindal, R. G. (1979). Distensibility of human pulmonary capillary blood vessels in the intraalveolar septa. *Microvascular Research,* **17,** 5–87.
Soivio, A., Nikinmaa, M., Nyholm, K. & Westman, K. (1981). The role of the gills in the responses of *Salmo gairdneri* to moderate hypoxia. *Comparative Biochemistry and Physiology,* **70A,** 133–9.
Steen, J. B. & Kruysse, A. (1964). The respiratory function of teleostean gills. *Comparative Biochemistry and Physiology,* **12,** 127–42.
Stevens, E. D. & Randall, D. J. (1967*a*). Changes in blood pressure, heart rate and breathing rate during moderate swimming activity in rainbow trout. *Journal of Experimental Biology,* **46,** 307–15.
Stevens, E. D. & Randall, D. J. (1967*b*). Changes in gas concentration in blood and water during moderate swimming activity in rainbow trout. *Journal of Experimental Biology,* **46,** 329–37.
Taylor, H. H. & Greenaway, P. (1979). The structure of the gills and lungs of the arid-zone crab, *Holthuisana* (*Austrothelphusa*) *transversa* (Brachyura: Sundathelphusidae) including observations on arterial vessels within the gills. *Journal of Zoology,* **189,** 359–84.
Vogel, W., Vogel, V. & Pfautsch, M. (1976). Arteriovenous anastomoses in rainbow trout gill filaments. *Cell and Tissue Research,* **167,** 373–85.
Vogel, W. O. P. (1978). Arteriovenous anastomoses in the afferent region of trout gill filaments (*Salmo gairdneri*). *Zoomorphologie,* **90,** 205–12.
Wood, C. M. (1974). A critical examination of the physical and adrenergic factors affecting blood flow through the gills of the rainbow trout. *Journal of Experimental Biology,* **60,** 241–65.
Wood, C. M. & Randall, D. J. (1973*a*). The influence of swimming activity on sodium balance in the rainbow trout (*Salmo gairdneri*). *Journal of Comparative Physiology,* **82,** 207–33.
Wood, C. M. & Randall, D. J. (1973*b*). The influence of swimming activity on water balance in the rainbow trout (*Salmo gairdneri*). *Journal of Comparative Physiology,* **83,** 257–76.
Wright, D. E. (1973). The structure of the gills of the elasmobranch, *Scyliorhinus canicula* (L.). *Zeitschrift für Zellforschung und Mikroskopische Anatomie,* **144,** 489–509.

LEON GOLDSTEIN

Gill nitrogen excretion

Branchial nitrogen excretion serves at least three functions. First, the gill is the major route of excretion of nitrogenous end-products of metabolism in most aquatic invertebrates and vertebrates. Second, in some species, especially those living in freshwater, nitrogen excretion is coupled, in part, to osmotic and ionic regulation. Third, foreign lipid-soluble nitrogenous compounds, taken in from the environment, are mainly excreted via the gills. This article deals mainly with the first function. The coupling of branchial nitrogen excretion to osmotic and ionic regulation will not be discussed at length since several fine reviews on this topic have been published in the past ten years and it has already been covered by Dr Evans in this symposium (*Salt and water exchange across vertebrate gills*). The limited number of studies which have been made on branchial excretion of foreign nitrogenous compounds are discussed briefly at the end of the article.

Although nitrogen excretion studies have been carried out on numerous aquatic species, in relatively few of these studies has gill nitrogen excretion been measured specifically. Often, total nitrogen excretion is measured with no way of knowing whether the nitrogen was excreted by the gills or via an extrabranchial route, e.g. the kidney. In many instances it is reasonable to assume that most of the nitrogen was excreted by the gills, but it is impossible to make a quantitative assessment of the exact amount of nitrogen eliminated by means of this route. Therefore, in preparing this article I have tried to select those studies in which gill excretion was measured separately. In some cases this was not possible, but the assumption that the measured excretion took place almost exclusively via the gills seemed reasonable.

Most studies in which gill nitrogen excretion has been measured separately have been done on fish. Therefore, this article deals mainly with nitrogen excretion in fish. However, the principles and processes derived from studies on this group should apply to branchial excretion in other aquatic organisms as well.

Chemical nature of excreted nitrogen

In fish, nitrogen excreted by the gills is mainly in the forms of ammonia and urea (Forster & Goldstein, 1969). Other nitrogenous end-products, such as trimethylamine oxide, are excreted in only minor amounts by the gills. Excretion of nitrogen as ammonia has the advantage that no expenditure of energy is required for the conversion of precursors of ammonia (e.g. amino acids) into ammonia. In contrast, formation of urea from nitrogenous precursors requires the expenditure of three moles of ATP per mole of urea synthesised. Therefore, most aquatic organisms excrete nitrogen in the form of ammonia (Delaunay, 1931). However, in some species and under some circumstances urea is the major nitrogenous end-product excreted by the gills.

In contrast to the teleost fishes, the chondrichthian fishes (sharks, rays, skates and chimera) and the coelacanth, *Latimeria chalumnae*, excrete the majority of their nitrogen via the gill in the form of urea (Forster & Goldstein, 1969). This is due to the fact that these marine fishes retain urea at a relatively high concentration (300 mmol l^{-1}) thereby maintaining osmotic equilibrium between their internal body fluids and the surrounding seawater (Smith, 1936). Some of this urea is excreted as part of the normal turnover of the nitrogenous end-product from the body fluids.

In some fish, such as the African lungfish, *Protopterus*, the nature of the nitrogenous end-product excreted by the gills is markedly affected by environmental conditions (Forster & Goldstein, 1969). When these fish are living in water, they excrete the majority of their waste nitrogen via the gills in the form of ammonia. When water is scarce they form cocoons in the muddy bottom of the lake or pond in which they live and undergo aestivation until the water supply becomes more plentiful. In this state nitrogen can no longer be excreted via the gills (or the kidneys) and ammonia is kept from reaching toxic levels in the fish by conversion to urea in the liver. The urea accumulates in the body fluids where it can reach high levels during prolonged periods of aestivation. Therefore, when the water returns and aestivation ends, urea, not ammonia, is the major nitrogenous end-product excreted by *Protopterus* until all the extra nitrogen accumulated during aestivation is eliminated (Smith, 1930).

In summary, ammonia is the major form in which nitrogen is excreted by the gills of aquatic organisms. However, under circumstances in which it becomes advantageous or necessary to retain nitrogenous end-products within the body, urea replaces ammonia, either permanently or temporarily, as the major form of nitrogen excreted by the gills.

Sources of nitrogen excreted by the fish gill

Although there is general agreement that urea excreted by the fish gill arises from that present in the blood circulating through the gills, controversy has surrounded the origin of branchially-excreted ammonia in the past. Homer Smith originally suggested that ammonia excreted by the fish gill originated from that present in the blood. Later, considering the fact that ammonia does not accumulate in the body fluids of aestivating lungfish, even though gill excretion ceases, he suggested that branchially-excreted ammonia must be formed in the gills (Smith, 1953).

More recently, experimental evidence has been obtained which supports the view that branchially-excreted ammonia is derived, for the most part, from ammonia in the blood circulating through the gills. In one experiment the extraction of ammonia from blood circulating through the gills was measured in the teleost *Myoxocephalus scorpius* and compared to the rate of branchial ammonia excretion in this fish (Goldstein, Forster & Fanelli, 1964). Branchial extraction of free amino acids, which were presumed to be the most logical precursors of ammonia formed within the branchial cells, was measured in the same experiment. As shown in Fig. 1, extraction of preformed ammonia from blood circulating through the gills accounted for about two-thirds of the excreted ammonia; the remainder could be accounted for by extraction of α-amino acid nitrogen from the plasma. Thus, the data indicate that in this marine teleost branchially excreted ammonia is derived mainly from blood ammonia and to a minor extent from ammonia formed in the gills by the enzymic deamination of α-amino acid nitrogen extracted from the blood by the gills. It may be noted that the fish gill has been shown to have the requisite enzymes for the deamination of α-amino acids (Goldstein & Forster, 1961) and other possible precursors of ammonia such as the purine bases (Makarewicz & Zydowo, 1962).

Results of experiments done on the freshwater teleost *Cyprinus carpio* (Pequin, 1962) indicate that, just as in the sculpin, ammonia excreted by the carp gill is derived mainly from that extracted by the gill from the branchial circulation. Thus, although the number of species investigated is limited, the current evi-

Fig. 1. Comparison of ammonia excretion and blood ammonia and amino acid extraction by the gills of the marine teleost, *Myoxocephalus scorpius*. Ammonia extracted from the blood can account for approximately two-thirds of that excreted by the gills. The remaining one-third could be formed by deamination of blood amino acids. (From Goldstein, Forster & Fanelli, 1964.)

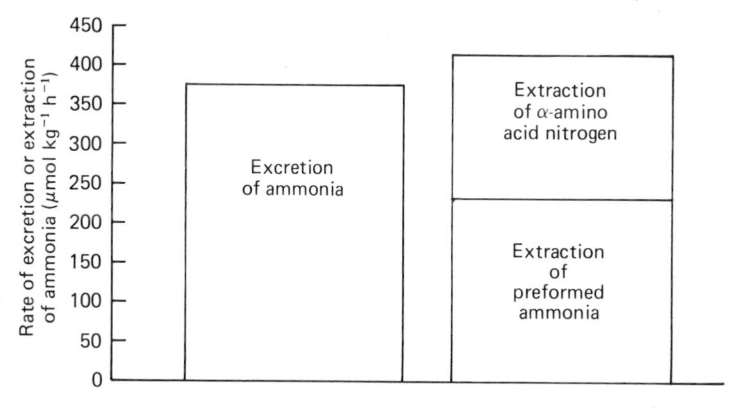

dence supports the view that the major fraction of ammonia excreted by the fish gill comes from the blood, while only a minor fraction originates from deamination processes within the branchial cells.

The mechanisms of the branchial excretion of ammonia and urea

Ammonia

Although numerous studies have been made on the mechanism of ammonia excretion by the gill, there is still considerable uncertainty as to the exact manner by which this process occurs. Part of the problem is related to the fact that ammonia exists in both the neutral and protonated forms in body fluids

$$NH_4^+ \rightleftharpoons NH_3 + H^+$$

The pK_a of the ammonium ion is about 9.2 at 25 °C; therefore between 1 and 10% of the ammonia molecules will be in the neutral form at blood pH. Since the early work of Jacobs (1940) demonstrating the effect of pH on the movement of acids and bases across cell membranes, it has been assumed that ammonia moves freely across biological membranes mainly in its neutral form, NH_3. It has also been assumed that the ammonium ion, NH_4^+, does not penetrate cell membranes easily and would have to be transported across membranes by some special process. Thus, it is possible for ammonia to be excreted across the gills by either non-ionic diffusion of free ammonia or by transport of ammonium ions. These processes are shown schematically in Fig. 2.

The early studies of Krogh (1939) suggested that the branchial excretion of ammonia was linked to cation absorption in freshwater fish. More recently, several investigators have examined the details of this exchange mechanism in the gills of both freshwater and saltwater fish and freshwater invertebrates (Shaw, 1960; Maetz & Garcia Romeu, 1964).

Considerable evidence has been accumulated in these studies to support Krogh's original idea that ammonium ion excretion is linked to cation absorption. For example, it has been shown that elevation of the concentration of ammonium ions in the blood stimulates the uptake of sodium ions by the gills of the goldfish *Carassius auratus* (Maetz & Garcia Romeu, 1964). However, few studies have been designed to test the degree of linkage between branchial ammonia excretion and cation absorption, so that it is not possible to tell exactly how much of the nitrogenous base is excreted by direct exchange with cations in the external environment and how much moves or is transported across the branchial cells independently.

In two freshwater species, the goldfish (*Carassius auratus*) and the crayfish (*Astacus pallipes*) there is excellent quantitative agreement between the rate of ammonia excretion and the uptake of sodium by the gills (Shaw, 1960; Maetz & Garcia Romeu, 1964) under normal conditions. However, if crayfish, carp or trout are placed in sodium-free media no diminution in the rate of ammonia excretion is observed (Shaw, 1960; de Vooys, 1968; Kerstetter, Kirschner & Rafuse, 1970). Other experiments done on teleost fish also cast doubt on the idea of a tight linkage between branchial ammonia excretion and cation absorption. For example, when goldfish are transferred from water at pH 7.2 to water at pH 6.1 sodium uptake decreases markedly but ammonia excretion remains unchanged (Maetz, 1973). In addition, in marine teleosts, in which the net movement of sodium ions across the gills is outward, the magnitude of branchial ammonia excretion is as great as that in freshwater fish (Maetz, 1972). It appears, therefore, that although there can be a close quantitative correlation between branchial ammonia excretion and the uptake of sodium ions under normal, basal conditions, experimental evidence indicates that this linkage can be loose and subject to physiological regulation. The question arises, therefore, as to the nature of the mechanism by which ammonia moves across the fish gill when this movement is not linked to the uptake of sodium ions.

It has been assumed that ammonia can readily cross the branchial cells in its neutral form. This assumption has been based in large part on experimental evidence derived from studies showing the effect of pH on the movement of ammonia across the cell membranes of erythrocytes (Jacobs, 1940) and renal tubules (Pitts, 1964), since pH-induced changes in the rate of ammonia movement are

Fig. 2. Schematic representation of possible mechanisms of ammonia movement across the gill. Ammonium ions could be transported across the branchial cells either in exchange for other cations (e.g. sodium) moving across the gill in the opposite direction, or accompanied by an anion (e.g. chloride or bicarbonate) moving in the same direction.

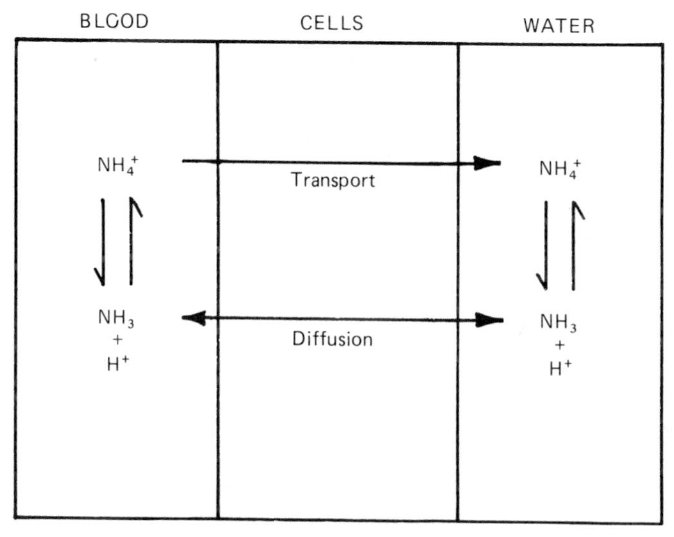

related to the effects of pH on the free ammonia/ammonium ion ratio in th
two systems. The ability of free ammonia to cross cell membranes more rea
than ammonium ions has been attributed to the greater solubility in lipids of
former. However, the hypothesis that ammonia moves across branchial me
branes as free ammonia has not been tested. In fact the assumption that f
ammonia is sufficiently soluble in lipids to allow it to move across the branc
membranes has not been examined. The effects of the lipid solubility of a se
of organic compounds on their rates of branchial excretion have been measu
in the spiny dogfish, *Squalus acanthias* (Maren, Embry & Broder, 1968).
shown in Fig. 3 a good correlation between lipid solubility, expressed as
partition coefficient between chloroform and water ($CHCl_3/H_2O$), and the
of branchial excretion was found over a wide range of lipid solubilities. W
the data for the chloroform/water partition coefficient for ammonia (Bell & Fe
1911) and the rate of branchial excretion of ammonia (Goldstein & Forster, 19
are plotted (Fig. 3), it is quite apparent that ammonia excretion by the gil
much greater than would be predicted from the lipid solubility of free ammo
This comparison, therefore, does not support the idea that the lipid solubility
free ammonia accounts for the rapid excretion of ammonia across the fish g
However, it could be argued that free ammonia could cross the branchial m
branes rapidly because of its small size and lack of charge in the neutral fo
In this connection a study on the mechanism of ammonia excretion in the
kidney indicated that the pH-dependent excretion of ammonia by the renal
bules is due to the diffusion of free ammonia through aqueous channels in
cell membrane (Bourke, Asatoor & Milne, 1972). We recently designed
experiment to test the hypothesis that ammonia is excreted by the fish as
ammonia.

The ratio of neutral/ionised forms of weak acids and bases in solution is de
mined by the pH of the solution and the p*K* of the acid or base as stated in
Henderson–Hasselbach equation

$$pH = pK_a + \log \text{[base]/[acid]}$$

In the case of ammonia the equation would be

$$pH = 9.2 + \log [NH_3]/[NH_4^+]$$

If ammonia moves across the gill mainly as free ammonia, i.e. the gill is assur
to be relatively impermeable to ammonium ions, then the rate of branc
ammonia excretion will be determined by the difference in the concentratio
free ammonia on the two sides (blood and water) of the gill

$$[NH_3]_{blood} - [NH_3]_{water}$$

If the gills are being perfused in the normal flow-through manner, the concentration of free ammonia on the water side of the gill will be negligible and the rate of excretion will be related only to the concentration of free ammonia in the blood.

Using the isolated fish head preparation (Payan, 1978) we perfused the gills of the sculpin, *Myoxocephalus octodecimspinosus* with fish Ringer's solution containing 1.0 mmol l^{-1} ammonium chloride. The concentration of free ammonia in the perfusion medium was varied by adjusting the pH of the medium to either 7.8 or 6.9. At the higher pH the calculated concentration of free ammonia

Fig. 3. Relation of lipid solubility to branchial clearance of a series of organic compounds (Maren, Embry & Broder, 1968) and ammonia (Goldstein & Forster, 1962) by the gills of the dogfish, *Squalus acanthias*. Each small dot shows the gill clearance and lipid solubility (expressed as the chloroform water partition coefficient, $CHCl_3/H_2O$) of a specific organic compound.

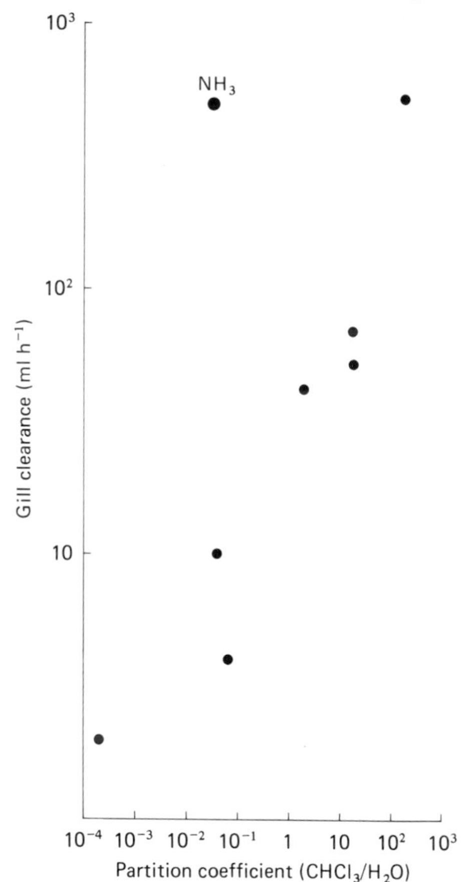

was 0.04 mmol l^{-1} whereas at the lower pH the calculated concentration was 0.005 mmol l^{-1}. In this preparation the gills are perfused with seawater which contains negligible amounts of free ammonia at the beginning and only small amounts, compared to the medium, at the end of the perfusion. Therefore, according to eqn (3), the rate of ammonia excretion by the gills should have been directly related to the concentration of free ammonia in the blood (perfusion medium): approximately eight times faster when the perfusion medium was at pH 7.8 than when it was at pH 6.9, if ammonia crossed the gills in the free form.

As shown in Fig. 4, the rate of ammonia excretion observed in preparations perfused at pH 7.8 was only slightly greater (approximately 30%) than that in preparations perfused at pH 6.9, and this difference was not statistically significant. Thus, the major factor determining the rate of ammonia excretion across the gills could not have been the concentration of free ammonia in the blood, and ammonia excretion is more likely related to processes involved in the movement of ammonium ions across the gills. Although we have only just begun to study the nature of these processes, it seems that an ammonium ion transport system may be involved in the excretion of ammonia by the gills.

Urea

As mentioned previously, urea is the chief nitrogenous end-product in chondrichthyian fishes and the coelacanth (*Latimeria chalumnae*) and is retained

Fig. 4. Lack of relation of branchial ammonia excretion to the concentration of free ammonia by the isolated head of the marine teleost, *Myoxocephalus octodecimspinosus*. The isolated head was perfused with medium containing 1.0 mmol l^{-1} ammonia ($NH_4^+ + NH_3$). The concentration of free ammonia was varied (0.04 mmol l^{-1} and 0.005 mmol l^{-1}) by adjusting the perfusion medium pH to 7.8 and 6.9 respectively. (From Goldstein, Claiborne & Evans 1982.)

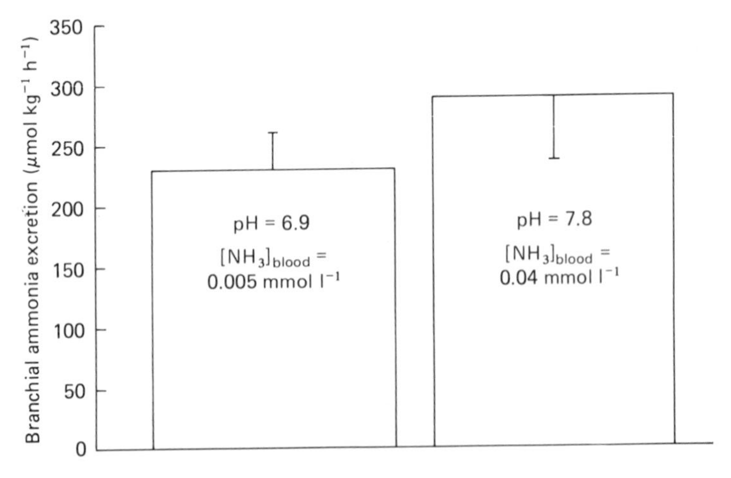

in these species for osmoregulatory purposes. Urea retention is the result of both active renal reabsorption (Forster, 1967) and relative impermeability of the gills (Boylan, 1967) of these fishes to urea. Although few experiments have been done in which gill permeability to urea has been measured directly, it is clear that most of the urea excreted by chondrichthyians leaves via the gills (Payan, Goldstein & Forster, 1973). Therefore, measurements of total body urea excretion carried out on fish maintained in seawater reflect, for the most part, what is happening at the gills. These measurements indicate that the gill is relatively impermeable to urea (Goldstein, Oppelt & Maren, 1968; Goldstein & Forster, 1971; Haywood, 1974). In fact, despite the huge chemical gradient that exists across the gill (300–400 mmol l^{-1}), only a small percentage of the urea circulating through the gills is lost per day.

The permeability of the dogfish (*Squalus acanthias*) gill to urea has been examined in detail (Boylan, 1967). In these experiments the dogfish were restrained in a divided chamber in which the gills could be perfused and the perfusate collected without contamination from surrounding seawater. Using radioisotopic tracers, measurements were made of the branchial excretion rates of [^{14}C]urea and [^{3}H]water. Gill filament area was measured by morphometric techniques. Branchial permeability coefficients were calculated from the data on rates of excretion and gill surface area and compared with the permeability coefficients of urea and water in another epithelial membrane, the toad urinary bladder. As seen in Table 1, the dogfish gill is a relatively tight epithelial membrane with regard to both urea and water permeabilities. Permeability of the dogfish gill to urea is only 1/30–1/40 of that of the toad bladder. Similarly, the permeability of the dogfish gill to water is one-tenth, or less, of that observed in the toad bladder. The relative impermeability of the dogfish gill membranes to urea contrasts sharply with the rapid rate of movement of this compound across other cell membranes in this fish (Fenstermacher, Sheldon, Ratner & Roomet, 1972).

The low permeability of the gill to urea may be solely due to physical factors, or there may be an active process in the branchial cells that prevents the loss of urea across the gills. For example, there might be a pump located on the baso-

Table 1. *Relative permeabilities of dogfish* (Squalus acanthias) *gill and toad* (Bufo marinus) *urinary bladder to urea and water*

	Permeabilities (cm s^{-1})			
	Urea	Water	Thiourea	Reference
Dogfish gill	8×10^{-8}	8×10^{-6}	8×10^{-8}	Boylan (1967)
Toad urinary bladder	3×10^{-6}	9×10^{-5}	—	Maffly & Leaf (1959)

lateral (blood) side of the cell which transports urea diffusing from the blood back into the extracellular fluid. If such a pump exists it would be expected to exhibit the characteristics of stereospecificity, temperature dependency and saturation kinetics as do other transport systems. Evidence has been sought for the presence of a urea-retaining transport system in the gills of the dogfish (Boylan, 1967). The rate of branchial excretion of thiourea (a sulphur analogue of urea) has been measured and compared with that of urea. The dogfish renal tubules are able to differentiate between urea and its sulphur analogue and reabsorb only about 35% of filtered thiourea compared with 95% or more of filtered urea (Boylan & Lockwood, 1962). However, Boylan & Lockwood found that gill permeability to thiourea and urea were approximately the same (Table 1).

Transport systems usually have high temperature coefficients. Boylan, Feldman & Antowiak (1963) measured the effects of temperature change (in the range of 1–30°C) on the rate of excretion of urea by the dogfish gill. Branchial urea excretion was constant between 1 and 15°C. Above 15°C gill excretion increased markedly. Boylan (1967) suggested that the marked increase in urea excretion above 15°C could have been due to a phase change in gill membrane structure, failure of the putative branchial urea pump or failure of some other unknown factor(s) contributing to membrane resistance. The excretion data obtained in the temperature range of 1–15°C are probably the most significant, since this encompasses the range of temperatures likely to be encountered by *Squalus acanthias,* i.e. the physiological temperature range. Failure to find a significant effect of temperature on gill urea retention over a 15°C range would argue strongly against the existence of a transport system for urea in the branchial cell membranes, since most transport systems show a high temperature dependency.

Attempts to demonstrate saturation kinetics for urea transport were unsuccessful (Boylan, Feldman & Antowiak, 1963). Intravenous loading of dogfish with exogenous urea produced results that were opposite to those expected from saturation of a transport system with substrate; the rate of branchial urea excretion increased exponentially as plasma urea concentration was raised above normal. For example, a three-fold increase in plasma urea concentration produced a twenty-fold rise in urea excretion. These results argue against the existence of a saturable, urea-retaining pump in the dogfish gill. As Boylan (1967) has pointed out, if elevation of the blood urea concentration saturated an inwardly directed urea pump, further elevation of the blood urea concentration should produce a linear, not an exponential, rise in branchial urea excretion. The exponential rise may have been the result of non-specific effects of a relatively high concentration of urea (e.g. disruption of hydrogen bonds in the cell membrane) on permeability.

Finally, injection of mercurial compounds known to be inhibitors of biological transport systems has no effect on the rate of urea excretion by the dogfish gill (Farber, Gerstein & Boylan, 1965). Thus, in a broad range of studies employing a wide variety of treatments and conditions, no evidence has been found to support the idea that the low permeability of the dogfish gill to urea is due to the operation of an inwardly-directed branchial urea pump. In the absence of such evidence, one is left with the conclusion that the gill membranes themselves form a rather tight physical barrier to the diffusion of urea.

Branchial excretion of foreign nitrogenous compounds

As the aqueous environment becomes more polluted with xenobiotics (foreign compounds), aquatic organisms are being exposed to, and are accumulating, a variety of foreign compounds, many of which are nitrogenous. Although the more polar organic nitrogenous compounds may be excreted by the kidney, renal excretion is limited for most foreign nitrogenous compounds in fish (Maren, Embry & Broder, 1968); they are excreted mainly by the gills. The excretion of a wide variety of foreign nitrogenous compounds has been studied in fishes. It has been found that the gills of fish are relatively impermeable to many foreign nitrogenous compounds. For example, antipyrine, a moderately lipid-soluble compound (the chloroform/water partition coefficient is 28) is cleared from the body fluids of the dogfish (*Squalus acanthias*) and sculpin (*Myoxocephalus scorpius*) via the gills at a rate of less than 5%/h (Rall, Bachur & Ratner, 1966).

The rate of movement of antipyrine across the dogfish gill is much slower than the rate of entry of this compound from the blood into the cerebrospinal fluid. The rate constant of the latter is 0.07 min^{-1} whereas the former rate constant is 0.003 min^{-1} (Adamson, 1967). Thus, foreign nitrogenous compounds do not cross branchial membranes readily, even if they are relatively lipid-soluble. If a foreign nitrogenous compound is very lipid-soluble, it is rapidly eliminated by the fish gill. Maren, Embry & Broder (1968) have shown that the fish anaesthetic tricaine methane sulphonate is rapidly excreted by the dogfish gill.

This highly lipid-soluble compound (the chloroform/water partition coefficient at pH 7.4 is 312) is cleared from the body fluids of the dogfish via the gills at a rate of almost 50% per hour. In fact its rate of excretion is limited only by gill blood flow. Thus, when the lipid solubility of a foreign nitrogenous compound reaches a critical level (i.e. when the chloroform/water partition coefficient is

C_6H_5 O N N CH_3 CH_3

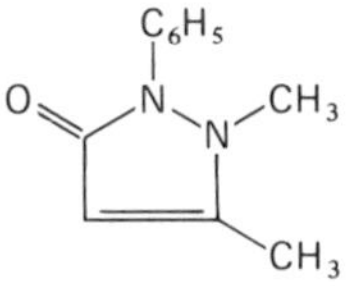

Antipyrine

$COOC_2H_5$

NH_2 · CH_3SO_3H

Tricaine methane sulphonate

somewhere between 28 and 312) the compound can be excreted rapidly by the fish gill. The question of why the fish gill is so impermeable to moderately lipid-soluble compounds such as antipyrine, as well as to small water-soluble molecules such as urea, is intriguing and should be explored.

References

Adamson, R. H. (1967). Drug metabolism in marine vertebrates. *Federation Proceedings,* **26,** 1047–54.

Bell, J. M. & Feild, A. L. (1911). The distribution of ammonia between water and chloroform. *Journal of the American Chemical Society,* **33,** 940–3.

Bourke, E., Asatoor, A. M. & Milne, M. D. (1972). Mechanisms of excretion of some low-molecular-weight bases in the rat. *Clinical Science,* **42,** 635–42.

Boylan, J. W. (1967). Gill permeability in *Squalus acanthias.* In *Sharks, Skates and Rays,* ed. P. W. Gilbert, R. F. Mathewson & D. P. Rall, pp. 197–206. Baltimore: Johns Hopkins Press.

Boylan, J. W., Feldman, B. & Antowiak, D. (1963). Factors affecting gill permeability in *Squalus acanthias. The Bulletin of the Mount Desert Island Biological Laboratory,* **5,** 29.

Boylan, J. W. & Lockwood, M. (1962). Urea and thiourea excretion by dogfish kidney and gill: effect of temperature. *The Bulletin of the Mount Desert Island Biological Laboratory,* **4,** 25.

Delaunay, H. (1931). L'excretion azotee des invertebres. *Biological Reviews,* **6,** 265–301.

Farber, J., Gerstein, B. & Boylan, J. W. (1965). Gill permeability to urea in *Squalus acanthias*: effect of various agents. *The Bulletin of the Mount Desert Island Biological Laboratory,* **5,** 14.

Fenstermacher, J., Sheldon, F., Ratner, J. & Roomet, A. (1972). The blood to tissue distribution of various polar materials in the dogfish, *Squalus acanthias. Comparative Biochemistry and Physiology,* **42A,** 195–204.

Forster, R. P. (1967). Osmoregulatory role of the kidney in cartilaginous fishes (*Chondrichthyes*). In *Sharks, Skates and Rays,* ed. P. W. Gilbert, R. F. Mathewson & D. P. Rall, pp. 187–95. Baltimore: Johns Hopkins Press.

Forster, R. P. & Goldstein, L. (1969). Formation of excretory products. In *Fish Physiology,* vol. 1, ed. W. S. Hoar & D. J. Randall, pp. 313–50. New York: Academic Press.

Goldstein, L., Claiborne, J. B. & Evans, D. H. (1982) Ammonia excretion by the gills of two marine fish: The importance of NH_4^+ permeance. *Journal of Experimental Zoology,* **219,** 395–7.

Goldstein, L. & Forster, R. P. (1961). Source of ammonia excreted by the gills of the marine teleost, *Myoxocephalus scorpius. American Journal of Physiology,* **200,** 1116–18.

Goldstein, L. & Forster, R. P. (1962). The relative importance of gills and kidneys in the excretion of ammonia and urea by the spiny dogfish (*Squalus acanthias*). *The Bulletin of the Mount Desert Island Biological Laboratory,* **4,** 33.

Goldstein, L. & Forster, R. P. (1971). Osmoregulation and urea metabolism in the little skate *Raja erinacea*. *American Journal of Physiology,* **220,** 742–6.

Goldstein, L., Forster, R. P. & Fanelli, G. M., Jr. (1964). Gill blood flow and ammonia excretion in the marine teleost, *Myoxocephalus scorpius*. *Comparative Biochemistry and Physiology,* **12,** 489–99.

Goldstein, L., Oppelt, W. W. & Maren, T. H. (1968). Osmotic regulation and urea metabolism in the lemon shark *Negaprion brevirostris*. *American Journal of Physiology,* **215,** 1493–7.

Haywood, G. P. (1974). The exchangeable ionic space, and salinity effects upon ion, water, and urea turnover rates in the dogfish *Poroderma africanum*. *Marine Biology,* **26,** 69–75.

Jacobs, M. H. (1940). Some aspects of cell permeability to weak electrolytes. *Cold Spring Harbor Symposia on Quantitative Biology,* **8,** 30–9.

Kerstetter, T. H., Kirschner, L. B. & Rafuse, D. D. (1970). On the mechanism of sodium ion transport by the irrigated gills of rainbow trout (*Salmo gairdneri*). *Journal of Experimental Biology,* **56,** 342–59.

Krogh, A. (1939). *Osmotic Regulation in Aquatic Animals*. London: Cambridge University Press.

Maetz, J. (1972). Interaction of salt and ammonia transport in aquatic organisms. In *Nitrogen Metabolism and the Environment,* ed. J. W. Campbell & L. Goldstein, pp. 105–54. New York: Academic Press.

Maetz, J. (1973). Na^+/NH_4^+, Na^+/H^+ exchanges and NH_3 movement across the gill of *Carassius auratus*. *Journal of Experimental Biology,* **58,** 255–75.

Maetz, J. & Garcia Romeu (1964). The mechanism of sodium and chloride uptake by the gills of the fresh-water fish, *Carassius auratus*. *Journal of General Physiology,* **47,** 1209–27.

Maffly, R. H. & Leaf, A. (1959). Potential of water in mammalian tissues. *Journal of General Physiology,* **42,** 1257–75.

Makarewicz, W. & Zydowo, M. (1962). Comparative studies on some ammonia producing enzymes in the excretory organs of vertebrates. *Comparative Biochemistry and Physiology,* **6,** 269–75.

Maren, T. H., Embry, R. & Broder, L. E. (1968). The excretion of drugs across the gill of the dogfish, *Squalus acanthias*. *Comparative Biochemistry and Physiology,* **226,** 853–64.

Payan, P. (1978). A study of the Na^+/NH_4^+ exchange across the gill of the perfused head of the trout (*Salmo gairdneri*). *Journal of Comparative Physiology,* **124,** 181–8.

Payan, P., Goldstein, L. & Forster, R. P. (1973). Gills and kidneys in ureosmotic regulation in euryhaline skates. *American Journal of Physiology,* **224,** 367–72.

Pequin, L. (1962). Les teneurs en azote ammoniacal du sang chez la carpe (*Cyprinus carpio L.*). *Comptes Rendus,* **255,** 1795–7.

Pitts, R. F. (1964). Renal production and excretion of ammonia. *American Journal of Medicine,* **36,** 720–41.

Rall, D. P., Bachur, N. R. & Ratner, J. H. (1966). The movement of foreign organic compounds across the gills of marine animals. *The Bulletin of the Mount Desert Island Biological Laboratory,* **6,** 31–2.

Shaw, J. (1960). The absorption of sodium ions by the crayfish *Astacus pallipes*

Lereboullet. III. The effect of other cations in the external solution. *Journal of Experimental Biology*, **37**, 548–56.
Smith, H. W. (1930). Metabolism of the lung-fish, *Protopterus aethiopicus*. *Journal of Biological Chemistry*, **88**, 97–130.
Smith, H. W. (1936). The retention and physiological role of urea in the elasmobranchii. *Biological Reviews*, **11**, 49–82.
Smith, H. W. (1953). *From Fish to Philosopher*. Boston: Little, Brown & Co.
de Vooys, G. G. N. (1968). Formation and excretion of ammonia in Teleostei. I. Excretion of ammonia through the gills. *Archives Internationales de Physiologie et de Biochemie*, **76**, 268–72.

J.C.RANKIN, R.M.STAGG & LIANA BOLIS

Effects of pollutants on gills

The effects of pollutants on aquatic organisms have recently been extensively reviewed in this series (Lockwood, 1976). The purpose of this chapter is to remind the reader of the particular vulnerability of the gills to toxic substances and to mention briefly a few examples to illustrate some recent lines of research.

Whilst the long-term effects of pollutants on animals may best be assessed at the level of their effects on populations in the field, and in particular on the reproductive success of these populations (Sprague, 1976), study of their mechanisms of action on organs such as gills can yield useful practical information. It is particularly important to be able to understand the ways in which different toxic substances and varying environmental conditions can interact in complex ways.

Reductions in respiratory efficiency may only be critical during occasional periods either of exposure to hypoxia or of increased metabolic activity. Substances which affect osmoregulation may only present the fish with serious problems during stressful osmoregulatory situations such as migration between freshwater and seawater. Laboratory toxicity tests carried out under carefully controlled conditions may be much more convenient than prolonged field observations, but their design can be improved by the use of knowledge of how the substances tested exert their toxic effects, for example in the development of more sensitive bioassays for product control and effluent monitoring purposes.

In contrast to the situation in laboratory experiments, animals such as fish can often avoid toxic concentrations in the wild (provided that they can detect them – some pollutants interfere with chemoreception) by swimming away, but this may not always be possible if the pollutant affects swimming ability. Effects on respiratory rate have long been recognised as sensitive indicators of sublethal actions of pollutants in fish (Belding, 1929). Oil dispersants cause a dose-dependent decrease in heart rate in the cunner (*Tautogolabrus adspersus*) which may impair swimming ability (Kiceniuk, Penrose & Squires, 1978). Detergent interferes with the vasodilatory action of catecholamines on the gill vasculature of both eels (*Anguilla anguilla*) and brown trout (*Salmo trutta*) (Bolis & Rankin, 1980), possibly limiting their scope for activity, and vanadate ions may have the same effect since they cause a pronounced vasoconstriction in perfused eel gills (Bell, Kelly & Sargent, 1979).

Most studies on the effects of pollutants on gills have concentrated on their actions on either respiration (Hughes, 1976) or water and salt balance (Lloyd & Swift, 1976). Damage to the gills is likely to affect both processes, and conclusions about the cause of death are likely to depend on which aspect of physiology has been studied. For example, the death of rainbow trout (*Salmo gairdneri*) following intermittent exposure to chlorine, which causes structural damage to the gills (Bass, Berry & Heath, 1977), was said to be due to a blockage of respiratory gas transfer which led to a large decrease in arterial oxygen tension (Bass & Heath, 1977). The structural damage to the gills of the bluegill (*Lepomis macrochirus*) caused by ozone, on the other hand, was found to cause a rapid decrease in serum osmolality, with death occurring when it had fallen by about 50 mOsmol kg^{-1} (Paller & Heidinger, 1980).

Since many xenobiotics, particularly organic compounds such as pesticides, herbicides and detergents, pass rapidly through the gills and contaminate other organs via the bloodstream, disturbances in respiratory or osmoregulatory functions need not necessarily be a consequence of gill damage. The gills frequently play an important role in the uptake of pollutants, if only because they represent such a large surface area for diffusion relative to the rest of the body surface (0.75 to 48 times as great (Gray, 1954)). Lipid-soluble compounds readily pass through the phospholipid bilayers of cell membranes and the bioconcentration factors in fish for a large number of organic compounds correlate closely with their *n*-octanol/water partition coefficients (Veith, DeFoe & Bergstedt, 1979). The gills may even transport toxic substances, for instance nitrite ions (Bath & Eddy, 1980), into the body.

This chapter will concentrate on examples of three types of pollutant which have been shown to affect the gills directly: lipophilic organic compounds, heavy metal ions and acids, although the role of the gills in the metabolism and elimination of xenobiotics must not be forgotten. In the vendace (*Coregonus albula*), for example, the activity of the enzyme UDP-glucuronosyl transferase (which is involved in the glucuronidation of lipophilic waste-products to make them water-soluble) in gill microsomes is twice that in liver microsomes (Lindström-Seppä, Koivusaari & Hänninen, 1981). Chloride cells may be involved in the excretion of organic molecules as well as salts (Masoni & Payan, 1974) and it has been suggested that an increase in the number of chloride cells following metal exposure might be a factor in the removal of toxic ions such as zinc (Matthiessen & Brafield, 1973) or cadmium (Calamari, Marchetti & Vailati, 1980).

Effects of lipophilic organic compounds

Much attention has been focused on detergents, which can pollute rivers and the sea from domestic sources (sewage) or when used as oil dispersants.

Exposure of fish to sodium lauryl sulphate (SLS) at a concentration of 100 mg l^{-1} results in a lifting of the upper layer of the lamellar epithelium and an invasion of the large subepithelial spaces thus created by lymphocytes and granulocytes (Abel & Skidmore, 1975). These spaces increase the diffusion distance between water and blood, but 100 mg l^{-1} of linear alkylate sulphonate (LAS) nevertheless caused a two-and-a-half-fold increase in the entry of water (measured as the influx of tritiated water) in perfused rainbow trout gills (Jackson & Fromm, 1977). The magnitude of the response was related to the detergent concentration, 10 mg l^{-1} producing a 50% increase. Surfactants might be expected to increase the permeability of cell membranes, and also to affect the layer of mucus which coats the gills.

Sublethal concentrations of oil dispersants and the water-soluble fraction of crude oil itself caused ultrastructural damage to the gills of the limpet (*Patella vulgata*); the initial action was thought to be on the outer cell membrane (Nuwayhid, Spencer, Davies & Elder, 1980). Oil dispersant caused a decrease in plasma sodium concentration in freshwater rainbow trout but an increase in that of fish acclimated to seawater (McKeown & March, 1978) and this could have been due to an increased permeability of the cell membranes. Such an effect has been suggested to explain the reversible increase in plasma sodium levels produced by lipid-soluble components of crude oil in seawater-adapted coho salmon (*Oncorhynchus kisutch*) (Morrow, Gritz & Kirton, 1975).

The effects of lipid-soluble pollutants may often be due to their ability to dissolve in, and affect the integrity of, cell membranes. The toxicity of a range of phenol derivatives and of DDT in rats has been shown to correlate with the degree of perturbation of membrane phospholipids as measured by differential scanning calorimetry of dipalmitoyl phosphatidylcholine liposomes (Packham, Thompson, Mayfield, Inniss & Kruuv, 1981). As well as affecting permeability, effects on membrane fluidity might be expected to influence the activities of membrane-bound enzymes. For example, the activity of the enzyme sodium- and potassium-activated ATPase in the gills of the rock crab (*Cancer irroratus*) was inhibited by DDT at a concentration of 0.5 mg l^{-1} and above *in vitro,* or a dose of 0.1 mg kg^{-1} *in vivo* (Neufeld & Pritchard, 1979). DDT (3 mg kg^{-1} $48h^{-1}$) also inhibited sodium- and potassium-activated ATPase in the gills of rainbow trout adapted to seawater and the degree of inhibition correlated well with increased serum osmolality and sodium concentration (Leadem, Campbell & Johnson, 1974). However, very low (25 μg l^{-1}) concentrations of DDT stimulated sodium uptake in isolated perfused gills of the freshwater carp (*Cyprinus carpio*) in contrast to the pesticide aldrin and the herbicides atrazine and trifluralin which depressed sodium uptake (McBride & Richards, 1975).

In eels and brown trout which had been exposed to 1mg l^{-1} LAS *in vivo*, the ability of catecholamines to stimulate β-adrenergic receptors to produce vasodilation in isolated perfused gills was inhibited (Bolis & Rankin, 1980). The receptor–hormone complex exerts its effect by stimulating the membrane-bound enzyme adenyl cyclase to produce cyclic AMP within the cells. Whilst the detergent could act by disrupting the structure of the protein molecules involved, the reversibility of the inhibition produced by very low concentrations (6 μg l^{-1}) of SLS (Stagg, Rankin & Bolis, 1981; Fig. 1) suggests a primary action on membrane fluidity.

Effects of heavy metal ions

Uptake of heavy metal ions by gills can be rapid and may result in accumulation to high concentrations. For example, methylmercury levels in the gills greatly exceeded those in all other organs 24 h after exposure of rainbow trout to the toxicant (Olson, Squibb & Cousins, 1978) and in gill tissue of the

Fig. 1. Effect of sodium lauryl sulphate (SLS) on the log dose–response curve for the effect of noradrenaline on flow (expressed as percentage of flow before the addition of hormone) through isolated perfused eel gills. Control curves were obtained before and after an experimental curve in the presence of SLS (6 μg l^{-1}). Values are means ± s.e. with $n = 6$. (From Stagg, Rankin & Bolis, 1981.)

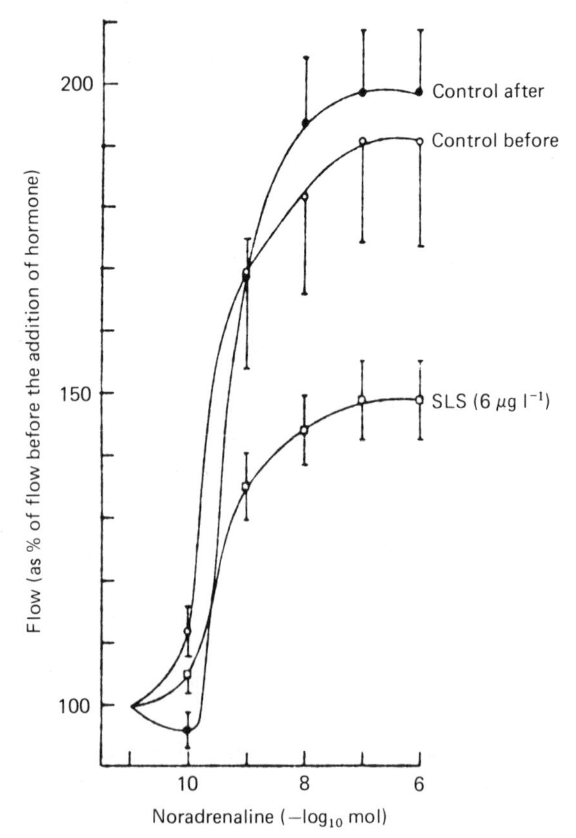

brook trout (*Salvelinus fontinalis*), cadmium can reach almost 1000 times the environmental concentration (Sangalang & Freeman, 1979). Dietary absorption may be the main route for heavy metal contamination (Bryan, 1976) but in crabs (*Carcinus maenas*) and shrimp (*Lysmata seticaudata*) absorption of zinc-65 from food appeared to be unimportant compared with direct uptake from the water (Renfro, Fowler, Heyraud & La Rosa, 1975). Uptake from solution is considered to occur by passive diffusion down gradients created by adsorption of the ions to the surfaces of cells and tissues. Obviously the rate of uptake will be dependent upon the concentration gradients (determined by both the binding capacity and external concentration of the metal) and the permeability of the body surface. The permeability of gills to heavy metal ions, as to other substances, may be affected by the external calcium ion concentration (Calamari, Marchetti & Vailati, 1980).

The effects of metal ions in fish respiration have been discussed by Hughes (1976) and many qualitative data on their effects on gill structure have been published. Zinc, for example, causes lifting of the outer layer of the lamellar epithelium and enlargement of the underlying lymph space (Skidmore & Tovell, 1972). Recent morphometric studies have quantified such effects and have shown that sublethal concentrations of nickel ions cause a reduction in the relative diffusing capacity of rainbow trout gills (Hughes, Perry & Brown, 1979).

Some heavy metals may disrupt the osmotic and ionic regulatory functions of the gill by affecting active ion transport and/or the permeability of the branchial epithelium. Many recent studies have related osmoregulatory perturbances to inhibition of branchial sodium- and potassium-activated ATPase (e.g. Stagg & Shuttleworth, 1982*a, b*). In the euryhaline flounder (*Platichthys flesus*), exposure to copper in the ambient water caused plasma electrolyte concentrations to increase in seawater-adapted fish and to decrease in freshwater-adapted fish (Table 1). Isolated gills from seawater-adapted flounders were perfused with, and bathed

Table 1. *Plasma ion concentrations in seawater-adapted and freshwater-adapted flounders* (Platichthys flesus) *after five to six weeks' exposure to low concentrations of copper. (Modified from Stagg & Shuttleworth, 1982a)*

	Seawater-adapted		Freshwater-adapted	
	Control	0.17 mg l^{-1} copper	Control	0.01 mg l^{-1} copper
Sodium	163.9 ± 1.5 (9)[a]	174.1 ± 3.2 (8)*	153.8 ± 2.1 (6)	132.2 ± 3.4 (7)*
Potassium	3.18 ± 0.05 (9)	3.52 ± 0.1 (8)	3.33 ± 0.18 (6)	2.77 ± 0.23 (7)
Chloride	141.8 ± 1.4 (9)	152.8 ± 3.0 (8)*	126.0 ± 1.4 (6)	112.7 ± 3.8 (7)*

[a] Values are means ± s.e. (*n*) given in mmol l^{-1}.

*$P < 0.05$ in all cases, for the comparison between control and copper-exposed fish.

in, an identical saline solution to eliminate diffusion potentials. The potential observed in such a situation is both dependent on metabolic activity (it is reduced by anoxia) and sensitive to the presence of ouabain (a specific inhibitor of sodium- and potassium-activated ATPase) supporting the suggestion that it is produced as a result of the electrogenic active transport of ions. Cupric ions in the perfusate produced a concentration-dependent inhibition of the potential, 100 μmol l^{-1} being virtually as effective as 10 μmol l^{-1} ouabain (Fig. 2*a*). The effect on the potential is closely paralleled by the inhibition *in vitro* of the sodium- and potassium-activated ATPase activity of the gill tissue (Fig. 2*b*). These results suggest that copper interferes with osmoregulation in the flounder by inhibiting branchial sodium- and potassium-activated ATPase. However despite the clear inhibition *in vitro,* sodium- and potassium-activated ATPase activity measured in gills of flounders exposed to copper in the ambient water was not significantly different from that in controls (Table 2). The use of a ouabain-binding technique to measure the number of sodium- and potassium-activated ATPase sites in the tissue showed that the fish responded to challenge with copper by increasing the number of enzyme units in order to maintain overall activity (Table 2). It has been suggested that ATPase activity in gills is modulated by changing the number of enzyme units (Sargent & Thompson, 1974). Branchial sodium- and potassium-activated ATPase activity is stimulated by cortisol (Epstein, Cynamon & McKay, 1971; Butler & Carmichael, 1972) and exposure of coho salmon to sublethal levels of copper resulted in elevated serum cortisol concentrations, although the more toxic metal cadmium did not produce this effect (Schreck & Lorz, 1978).

Table 2. *Effects of in vivo exposure to copper (200 μg l^{-1} for 42 days) on sodium- and potassium-activated ATPase and ouabain binding in the gills of seawater-adapted* Platichthys flesus. (*Modified from Stagg & Shuttleworth, 1982b*)

	Controls	Copper-exposed
Na^+- + K^+-activated ATPase (μmol P_i mg protein^{-1} 30 min^{-1})	6.75 ± 0.35 (8)[a]	8.04 ± 0.73 (6)[ns]
Ouabain binding (pmol mg dry^{-1})	15.62 ± 1.00 (11)	23.37 ± 1.58 (6)*

[a]Values are means ± s.e. (*n*).
ns, not significantly different.
*$P < 0.001$ for the comparison between control and copper-exposed fish.

The increase in the number of enzyme units (ouabain-binding sites) observed in the seawater-adapted flounder exposed to copper could conceivably have been due to increased cortisol secretion. This would result in similar sodium- and potassium-activated ATPase levels in control and copper-exposed fish despite the inhibition of the enzyme by copper.

In smolts of coho salmon, chromate (Sugatt, 1980) and copper (Lorz & McPherson, 1976) did not affect osmoregulation or survival in freshwater, although copper inhibited branchial sodium- and potassium-activated ATPase activity. However, osmoregulatory disturbances and high mortalities followed transfer to seawater, so exposure to the metal impaired the ability of the fish to cope with a subsequent osmoregulatory stress. In addition, very low copper concentrations (down to 5 μg l^{-1}) decreased the tendency to migrate downstream towards saltwater, so it is possible that the fish try to avoid osmotic stress if the water is polluted.

Inhibition of branchial sodium- and potassium-activated ATPase activity may not always explain heavy metal ion interference with osmoregulation. In rainbow trout mercuric and methylmercuric chlorides only inhibit the enzyme at lethal concentrations, whereas even sublethal levels decreased plasma sodium and chloride concentrations in freshwater-adapted fish and increased osmotic water uptake by isolated incubated gills; the primary effect of low concentrations

Fig. 2. (*a*) Effects of copper and ouabain (following copper) in the perfusate on potentials in isolated perfused gills from seawater-adapted *Platichthys* bathed and perfused with an identical saline solution. Values are means ± s.e., with *n* given in parentheses. (*b*) Effects of exposure to copper *in vitro* on the activity of sodium- and potassium-activated ATPase in gills from seawater-adapted *Platichthys*. (Activity measured as μmol inorganic phosphate produced mg protein^{-1} h^{-1}). Values are means ± s.e., with *n* given in parentheses. (Adapted from Stagg & Shuttleworth, 1982*b*.)

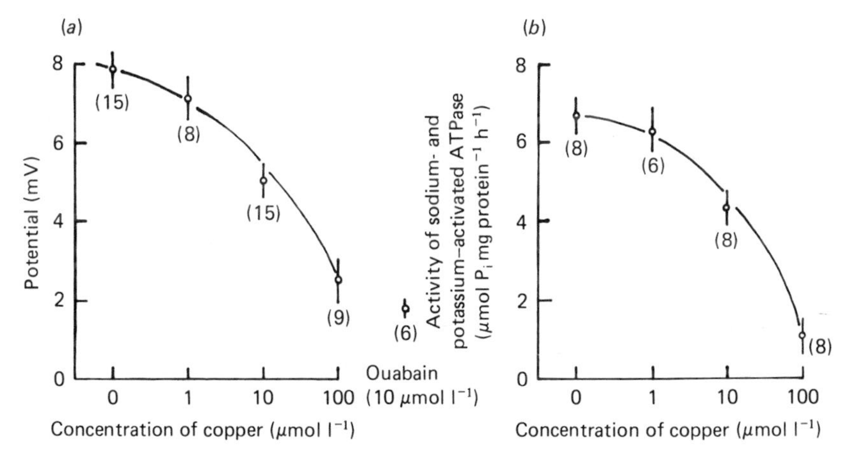

appeared to be to increase the permeability of the gill membrane (Lock, Cruijsen & van Overbeeke, 1981).

Various ATPases in microsomal preparations from rainbow trout gills, in addition to sodium- and potassium-activated ATPase, are inhibited by zinc (Watson & Beamish, 1981). Calcium-activated ATPase (which is thought to be involved in calcium excretion) in the gills of the roach, *Rutilis rutilis* was inhibited by copper, lead, zinc and mercury at concentrations of below 10 μmol l^{-1}. However, as with sodium- and potassium-activated ATPase from the flounder, prolonged exposure of fish to very low copper concentrations (<0.2 μmol l^{-1}) appeared to induce the formation of new enzyme units, presumably to maintain a constant level of calcium-activated ATPase in the gills (Shephard & Simkiss, 1978).

The ways in which fish adjust their physiology to counter the effects of toxicants are poorly understood. However it is clear that in the case of copper there are compensatory responses to the effects observed on both the calcium-activated and sodium- and potassium-activated ATPases. Furthermore in the flounder, evidence of adaptation at the biochemical level to the polluted situation has been obtained from a study of the effects of copper (*in vitro*) on the sodium- and potassium-activated ATPase from control and copper-exposed fish. The results suggest that the enzyme from copper-exposed fish is less sensitive to copper applied *in vitro*, implying that the enzyme itself has become modified in response to the presence of copper (Stagg & Shuttleworth, 1982*b*).

In conclusion, there are obviously a number of ways in which metal ions can interfere with osmoregulatory effector organs and their control mechanisms. However, much of the basic information about the normal physiology of fish is still lacking and in only a few cases have we begun to understand the interactions between the metal and the physiological processes.

Effects of acids

Organic acids of natural origin, such as dehydroabietic acid released into rivers by wood-pulp mills, may have specific toxic effects on fish (Iwama, Greer & Larkin, 1976), but this section is concerned solely with the effects of low pH caused by inorganic acids, principally sulphuric (originating mainly from coal burning) and nitric (originating mainly from petroleum burning) acids.

The effects of low pH on fish gill structure and function have been summarised in the review by Fromm (1980). Norwegian brown trout, suddenly exposed to pH values as low as 4.0 when melting snow rapidly released acid which had accumulated over the winter, showed reduced plasma sodium and chloride concentrations, the largest reductions being found in fish from areas where the greatest mortalities were experienced (Leivestad & Muniz, 1976). Brown trout gills are very permeable to protons which enter and produce a positive internal electrical potential, thereby causing an increased passive efflux of sodium ions (McWilliams & Potts, 1978; McDonald & Wood, 1981). The permeability to both protons and sodium ions was inversely related to the external calcium ion concentration, so effects of low pH are most severe in soft waters, where the buffering capacity is, in any case, very low. Trout acclimated to low pH are able to reduce sodium loss by decreasing the permeability of their gills to sodium ions (McWilliams, 1980*b*). Sodium uptake is also reduced on acute exposure to acid, although it returns to normal in a few days (McWilliams, 1980*a*).

Exposure to acid leads to a decrease in blood pH (which reduces the oxygen-carrying capacity of the blood), and, in severe cases, to reduced branchial oxygen uptake (Packer, 1979). Accumulation of mucus increases the diffusion distance for oxygen and may reduce the water flow rate between the secondary lamellae (Ultsch & Gros, 1979) or structural damage to the lamellar epithelium may occur (Fromm, 1980).

Conclusion

The selected examples given in this chapter will, it is hoped, give some idea of the interest inherent in the elucidation of the mechanisms of action of pollutants on gills, and indeed on other organs. The standard way of measuring the harm done by toxic substances is to determine the concentration which kills 50% of a group of animals (the LC_{50}). A 'safe' level of the substance can then be set at some fraction of this concentration. At the other extreme, 'safe' levels can be defined as those which do not affect the reproductive success of a wild population, but this is so difficult to quantify, in view of normal fluctuations in populations and variations in a multitude of interrelated environmental factors, that the importance of a particular pollutant could easily be overlooked. Both approaches are necessary, but there is also a need for a third approach – the study of the structural, physiological and behavioural changes produced by sublethal concentrations of toxicants – partly to act as a warning of unforeseen dangers and partly because these studies can lead to a greater understanding of basic biological mechanisms.

The survival of the human race depends in no small measure on our ability to foresee and protect ourselves from the environmental changes produced by an ever-expanding technology. Aquatic animals, with their necessity to expose a large area of delicate tissues, the gills, to the environment, can serve as sensitive indicators of the quality of the water which covers the majority of the surface of our earth and is so essential to all life.

We are grateful to Dr A. Oikari for helpful comments.

References

Abel, P. D. & Skidmore, J. F. (1975). Toxic effects of an anionic detergent on the gills of rainbow trout. *Water Research,* **9,** 759–65.

Bass, M. L., Berry, C. R. & Heath, A. G. (1977). Histopathological effects of intermittent chlorine exposure on bluegill (*Lepomis macrochirus*) and rainbow trout. *Water Research,* **11,** 731–5.

Bass, M. L. & Heath, A. G. (1977). Cardiovascular and respiratory changes in rainbow trout, *Salmo gairdneri,* exposed intermittently to chlorine. *Water Research,* **11,** 497–502.

Bath, R. N. & Eddy, F. B. (1980). Transport of nitrite ions across fish (*Salmo gairdneri*) gills. *Journal of Experimental Zoology,* **214,** 119–21.

Belding, D. L. (1929). The respiratory movements of fish as an indicator of toxic environment. *Transactions of the American Fisheries Society,* **59,** 238–45.

Bell, M. V., Kelly, K. F. & Sargent, J. R. (1979). Sodium orthovanadate, a powerful vasoconstrictor in the gills of the common eel, *Anguilla anguilla. Journal of the Marine Biological Association of the United Kingdom,* **59,** 429–35.

Bolis, L. & Rankin, J. C. (1980). Interactions between vascular actions of detergent and catecholamines in perfused gills of European eel, *Anguilla anguilla* L. and brown trout, *Salmo trutta* L. *Journal of Fish Biology,* **16,** 61–73.

Bryan, G. W. (1976). Some aspects of heavy metal tolerance in aquatic organisms. In *Effects of Pollutants on Aquatic Organisms,* ed. A. P. M. Lockwood, pp. 7–34. Cambridge University Press.

Butler, D. G. & Carmichael, F. J. (1972). (Na^+–K^+)-ATPase activity in eel (*Anguilla rostrata*) gills in relation to changes in environmental salinity: role of adrenocortical steroids. *General and Comparative Endocrinology,* **19,** 421–7.

Calamari, D., Marchetti, R. & Vailati, G. (1980). Influence of water hardness on cadmium toxicity to *Salmo gairdneri. Water Research,* **14,** 1421–6.

Epstein, F. H., Cynamon, M. & McKay, W. (1971). Endocrine control of Na–K-ATPase and seawater adaptation in *Anguilla rostrata. General and Comparative Endocrinology,* **16,** 323–8.

Fromm, P. O. (1980). A review of some physiological and toxicological responses of freshwater fish to acid stress. *Environmental Biology of Fishes,* **5,** 79–93.

Gray, I. E. (1954). Comparative study of the gill area of marine fishes. *Biological Bulletin of the Marine Biological Laboratory, Woods Hole,* **107,** 219–25.

Hughes, G. M. (1976). Polluted fish respiratory physiology. In *Effects of Pollutants on Aquatic Organisms,* ed. A. P. M. Lockwood, pp. 163–83. Cambridge University Press.

Hughes, G. M., Perry, S. F. & Brown, V. M. (1979). A morphometric study of effects of nickel, chromium and cadmium on the secondary lamellae of rainbow trout gills. *Water Research,* **13,** 665–79.

Iwama, G. K., Greer, G. L. & Larkin, P. A. (1976). Changes in some hematological characteristics of coho salmon (*Oncorhynchus kisutch*) in response to acute exposure to dehydroabietic acid (DHAA) at different exercise levels. *Journal of the Fisheries Research Board of Canada,* **33,** 285–9.

Jackson, W. F. & Fromm, P. O. (1977). Effect of a detergent on flux of tritiated water into isolated perfused gills of rainbow trout. *Comparative Biochemistry and Physiology,* **58C,** 167–71.

Kiceniuk, J. W., Penrose, W. R. & Squires, W. R. (1978). Oil spill dispersants cause bradycardia in a marine fish. *Marine Pollution Bulletin,* **9,** 42–5.

Leadem, T. P., Campbell, R. D. & Johnson, D. W. (1974). Osmoregulatory responses to DDT and varying salinities in *Salmo gairdneri* – I. Gill Na-K-ATPase. *Comparative Biochemistry and Physiology,* **49A,** 197–205.

Leivestad, H. & Muniz, I. P. (1976). Fish kill at low pH in a Norwegian river. *Nature,* **259,** 391–2.

Lindström-Seppä, P., Koivusaari, U. & Hänninen, O. (1981). Metabolism of xenobiotics by Vendace (*Coregonus albula*). *Comparative Biochemistry and Physiology,* **68C,** 121–6.

Lloyd, R. & Swift, D. J. (1976). Some physiological responses by freshwater fish to low dissolved oxygen, high carbon dioxide, ammonia and phenol with particular reference to water balance. In *Effects of Pollutants on Aquatic Organisms,* ed. A. P. M. Lockwood, pp. 47–69. Cambridge University Press.

Lock, R. A. C., Cruijsen, P. M. J. M. & van Overbeeke, A. P. (1981). Effects of mercuric chloride and methylmercuric chloride on the osmotic function of the gills in rainbow trout, *Salmo gairdneri* Richardson. *Comparative Biochemistry and Physiology,* **68C,** 151–9.

Lockwood, A. P. M. (1976). *Effects of Pollutants on Aquatic Organisms,* Society for Experimental Biology, Seminar Series, volume 2. Cambridge University Press.

Lorz, H. W. & McPherson, B. P. (1976). Effects of copper or zinc in fresh water on the adaptation to sea water and ATPase activity, and the effects of copper on migratory disposition of coho salmon (*Oncorhynchus kisutch*). *Journal of the Fisheries Research Board of Canada,* **33,** 2023–30.

McBride, R. K. & Richards, B. D. (1975). The effects of some herbicides and pesticides on sodium uptake by isolated perfused gills from the carp *Cyprinus carpio*. *Comparative Biochemistry and Physiology,* **51C,** 105–9.

McDonald, D. G. & Wood, C. M. (1981). Branchial and renal acid and ion fluxes in the rainbow trout, *Salmo gairdneri,* at low environmental pH. *Journal of Experimental Biology,* **93,** 101–18.

McKeown, B. A. & March, G. L. (1978). The effect of Bunker C oil and an oil dispersant on serum glucose, serum sodium and gill morphology in both freshwater and seawater acclimated rainbow trout, *Salmo gairdneri*. *Water Research,* **12,** 157–63.

McWilliams, P. G. (1980*a*). Effects of pH on sodium uptake in Norwegian brown trout (*Salmo trutta*) from an acid river. *Journal of Experimental Biology,* **88,** 259–67.

McWilliams, P. G. (1980*b*). Acclimation to an acid medium in the brown trout, *Salmo trutta*. *Journal of Experimental Biology,* **88,** 269–80.

McWilliams, P. G. & Potts, W. T. W. (1978). The effects of pH and calcium concentrations on gill potentials in the brown trout, *Salmo trutta*. *Journal of Comparative Physiology,* **126,** 277–86.

Masoni, M. & Payan, P. (1974). Urea, inulin and para-amino-hippuric acid excretion by the gills of the eel, *Anguilla anguilla* L. *Comparative Biochemistry and Physiology,* **47A,** 1241–4.

Matthiessen, P. & Brafield, A. E. (1973). The effects of dissolved zinc on gills of the stickleback *Gasterosteus aculeatus*. *Journal of Fish Biology,* **5,** 607–13.

Morrow, J. E., Gritz, R. L. & Kirton, M. P. (1975). Effects of some components of crude oil on young coho salmon. *Copeia,* **1975,** 326–31.

Neufeld, G. J. & Pritchard, J. B. (1979). Osmoregulation and gill Na,K-ATPase in the rock crab, *Cancer irroratus*: response to DDT. *Comparative Biochemistry and Physiology,* **62C,** 165–72.

Nuwayhid, M. A., Spencer Davies, P. & Elder, H. Y. (1980). Changes in the ultrastructure of the gill epithelium of *Patella vulgata* after exposure to North Sea crude oil and dispersant. *Journal of the Marine Biological Association of the United Kingdom,* **60,** 439–48.

Olson, K. R., Squibb, K. S. & Cousins, R. J. (1978). Tissue uptake, subcellular distribution and metabolism of $^{14}CH_3HgCl$ and $CH_3{}^{203}HgCl$ by rainbow trout, *Salmo gairdneri. Journal of the Fisheries Research Board of Canada,* **35,** 381–90.

Packer, R. P. (1979). Acid–base balance and gas exchange in brook trout (*Salvelinus fontinalis*) exposed to acidic environments. *Journal of Experimental Biology,* **79,** 127–34.

Packham, E. D., Thompson, J. E., Mayfield, C. I., Inniss, W. E. & Kruuv, J. (1981). Perturbation of lipid membranes by organic pollutants. *Archives of Environmental Contamination and Toxicology,* **10,** 347–56.

Paller, M. H. & Heidinger, R. C. (1980). Mechanism of delayed ozone toxicity to bluegill, *Lepomis macrochirus. Environmental Pollution* (*Series A*), **22,** 229–40.

Renfro, W. C., Fowler, S. W., Heyraud, M. & La Rosa, J. (1975). Relative importance of food and water in long-term zinc-65 accumulation of marine biota. *Journal of the Fisheries Research Board of Canada,* **32,** 1339–46.

Sangalang, G. B. & Freeman, H. C. (1979). Tissue uptake of cadmium in brook trout during chronic sublethal exposure. *Archives of Environmental Contamination and Toxicology,* **8,** 77–84.

Sargent, J. R. & Thompson, A. J. (1974). The nature and properties of the inducible sodium-plus-potassium ion dependent adenosine triphosphatase in the gills of eels (*Anguilla anguilla*) adapted to freshwater and seawater. *Biochemical Journal,* **144,** 69–75.

Schreck, C. B. & Lorz, H. W. (1978). Stress response of coho salmon (*Oncorhynchus kisutch*) elicited by cadmium and copper and potential use of cortisol as an indicator of stress. *Journal of the Fisheries Research Board of Canada,* **35,** 1124–9.

Shephard, K. & Simkiss, K. (1978). The effects of heavy metal ions on Ca^{2+}ATPase extracted from fish gills. *Comparative Biochemistry and Physiology,* **61B,** 69–72.

Skidmore, J. F. & Tovell, P. W. A. (1972). Toxic effects of zinc sulphate on the gills of rainbow trout. *Water Research,* **6,** 217–30.

Sprague, J. B. (1976). Current status of sublethal tests of pollutants on aquatic organisms. *Journal of the Fisheries Research Board of Canada,* **33,** 1988–92.

Stagg, R. M., Rankin, J. C. & Bolis, L. (1981). Effect of detergent on vascular responses to noradrenaline in isolated perfused gills of the eel, *Anguilla anguilla* L. *Environmental Pollution* (*Series A*), **24,** 31–7.

Stagg, R. M. & Shuttleworth, T. J. (1982*a*). The accumulation of copper in *Platichthys flesus* L. and its effects on plasma electrolyte concentrations. *Journal of Fish Biology,* **20,** 491–500.

Stagg, R. M. & Shuttleworth, T. J. (1982*b*). The effects of copper on ionic regulation by the gills of seawater-adapted flounder (*Platichthys flesus* L.) *Journal of Comparative Physiology,* in press.

Sugatt, R. H. (1980). Effects of sublethal sodium dichromate exposure in freshwater on the salinity tolerance and serum osmolality of juvenile coho salmon, *Oncorhynchus kisutch,* in seawater. *Archives of Environmental Contamination and Toxicology,* **9,** 41–52.

Ultsch, G. R. & Gros, G. (1979). Mucus as a diffusion barrier to oxygen: possible role in oxygen uptake at low pH in carp (*Cyprinus carpio*) gills. *Comparative Biochemistry and Physiology,* **62A,** 685–9.

Veith, G. D., DeFoe, D. L. & Bergstedt, B. V. (1979). Measuring and estimating the bioconcentration factor of chemicals in fish. *Journal of the Fisheries Research Board of Canada,* **36,** 1040–8.

Watson, T. A. & Beamish, F. W. H. (1981). The effects of zinc on branchial adenosine triphosphatase enzymes *in vitro* from rainbow trout, *Salmo gairdneri. Comparative Biochemistry and Physiology,* **68C,** 167–73.

INDEX